69th Northeastern Weed Science Society Annual Meeting 2015

Proceedings – Volume 69

Williamsburg, Virginia, USA
5 – 8 January 2015

Editor:

Arthur E. Gover

ISBN: 978-1-5108-3340-1

Printed from e-media with permission by:

Curran Associates, Inc.
57 Morehouse Lane
Red Hook, NY 12571

Some format issues inherent in the e-media version may also appear in this print version.

Printed by Curran Associates, Inc. (2017)

Northeastern Weed Science Society (NEWSS)
P.O. Box 25
Woodstown, NJ 08098
USA

Phone: 315-209-7580

northeasternweedscience@hotmail.com

Additional copies of this publication are available from:

Curran Associates, Inc.
57 Morehouse Lane
Red Hook, NY 12571 USA
Phone: 845-758-0400
Fax: 845-758-2633
Email: curran@proceedings.com
Web: www.proceedings.com

Proceedings

of the

Sixty-ninth Annual Meeting

of the

Northeastern Weed Science Society

Arthur E. Gover, Editor

NORTHEASTERN WEED SCIENCE SOCIETY

2015 Sustaining Members

Platinum

BASF The Chemical Company

syngenta

Gold

FMC

Dow AgroSciences

VALENT

Bayer CropScience

MONSANTO

Silver

BAAR Scientific LLC

ohp Partners with solutions

Gowan USA The Go To Company

DU PONT

UPI

pbi/gordon corporation An Employee-Owned Company

Bronze

Gylling Data Management
TeeJet® Technologies
Weeds, Inc.
Crop Management Strategies
LABSerivices

Table of Contents

Number in parentheses is the title number. Authors and keywords are indexed by title number.

NEWSS EXECUTIVE COMMITTEE - OFFICERS

President	**Gregory Armel** BASF Corporation 26 Davis Drive RTP, NC 27709 gregory.armel@basf.com
President-elect	**Rakesh Chandran** West Virginia University 1076 Agricultural Sciences Building PO Box 6108 Morgantown, WV, 26506 rschandran@mail.wvu.edu
Vice President	**Shawn Askew** Virginia Tech 435 Old Glade Rd. Blacksburg, VA 26505 saskew@vt.edu
Treasurer	**Erin Hitchner** Syngenta Crop Protection 380 Jefferson Road Elmer, NJ 08318 erin.hitchner@syngenta.com
Past President	**Dwight Lingenfelter** Penn State University 116 ASI Building University Park, PA, 16802 dxl18@psu.edu

NEWSS EXECUTIVE COMMITTEE - MEMBERS

Membership Chair	**Keith Burnell** Syngenta Crop Protection, LLC keith.burnell@syngenta.com
Editor	**Art Gover** Penn State University aeg2@psu.edu
Public Relations	**Todd Mervosh** Private Consultant tmervosh@att.net
Research & Education Coordinator	**Bryan Dillehay** Monsanto Corporation bryan.dillehay@monsanto.com
Sustaining Membership	**John O'Barr** BASF Corporation john.obarr@basf.com
CAST Representative	**Randy Prostak** University of Massachusetts rprostak@umext.umass.edu
Graduate Student Representative	**John Orlowski** University of Kentucky john.orlowski@uky.edu
WSSA Representative	**Prasanta Bhowmik** University of Massachuesetts pbhowmik@umass.edu
Science Policy Director	**Lee Van Wychen** Weed Science Society of America lee.vanwychen@wssa.net

NEWSS 2015 MEETING PROGRAM COMMITTEES

Program Committee	Chair:	Rakesh Chandran, West Virginia University
		Shawn Askew, Virginia Tech
Agronomy	Chair:	Bryan Dillehay, Monsanto
	Chair-elect:	Wes Everman, North Carolina State University
Ornamentals	Chair:	Jenny Kao-Kniffin, Cornell
	Chair-elect:	Todd Mervosh
Research Posters	Chair:	Aman Rana, Virginia Tech
	Chair-elect:	Barb Scott, University of Delaware
Turfgrass and Plant Growth Regulators	Chair:	Matthew Elmore, University of Tennessee
	Chair-elect:	Katelyn Venner, Virginia Tech
Vegetables and Fruit	Chair:	Jennifer D'Appollonio-Cote, University of Maine
	Chair-elect:	Katherine Ghantous, Univ. of Massachuesetts
Vegetation Management and Restoration	Chair:	Javier Vargas, University of Tennessee
	Chair-elect:	Brett Hartis, North Carolina State University
Weed Biology and Ecology	Chair:	Katherine Ghantous, Univ. of Massachuesetts
	Chair-elect:	Dan Tekiela, Virginia Tech

NEWSS COMMITTEES

Student Paper Contest	Chair:	Dwight Lingenfelter, Penn State
		Toni DiTommaso, Cornell University
		Mark VanGessel, University of Delaware
		Hilary Sandler, University of Massachuesetts
		Dave Yarborough, University of Maine
Student Poster Contest	Chair:	Keith Burnell, Syngenta
		Don Ganske, DuPont
		Todd Mervosh
		Art Gover, Penn State University
		Joseph Argentine, AmVac Chemical

NEWSS COMMITTEES (continued)

Photo Judging Contest	Chair:	Javier Vargas, University of Tennessee
		Erin Hitchner, Syngenta
		Rakesh Chandran, West Virginia University
		Art Gover, Penn State University
		Victoria Kleczewski, DuPont
		Hilary Sandler, University of Massachuesetts
Site Selection Committee:	Chair:	Rakesh Chandran, West Virginia University
		Greg Armel, BASF
		Erin Hitchner, Syngenta
		Keith Burnell, Syngenta
Awards Committee:	Chair	Dwight Lingenfelter, Penn State University
		Antonio DiTommaso, Cornell University
		Mark VanGessel, University of Delaware
		Hilary Sandler, University of Massachuesetts
		David Yarborough, University of Maine
Nomination Committee:	Chair	Antonio DiTommaso, Cornell University
		Javier Vargas, University of Tennessee
		Jon Johnson, Penn State University
		Todd Davis, Delaware Department of Agriculture
		Rob Richardson, North Carolina State University
Resolutions Committee:	Chair	Andy Senesac, Cornell University
		Courtney Stokes, Cornell University
		Jennifer D'Appollonio-Cote, Univeristy of Maine
Collegiate Weed Contest:	Chair:	Rakesh Chandran, West Virginia University
		Dwight Lingenfelter, Penn State University
		John Grande, Rutgers University
		Greg Armel, BASF
		John Orlowski, University of Kentucky
		Shawn Askew, Virginia Tech
		Wes Everman, North Carolina State University
Audit Committee		Erin Hitchner, Syngenta
		Todd Mervosh
		Jim Steffel, Lab Services, Inc.

NEWSS COMMITTEES (continued)

Archives Committee:	Chair:	Dan Kunkel, IR-4
		Dwight Lingenfelter,
Weed Science Field Days:	Chair:	Bryan Dillehay, Monsanto
		Quintin Johnson, University of Delaware
Past Presidents Committee:	Chair:	Roy Johnson, Waldrum Specialties
		Dwight Lingenfelter, Penn State University
Herbicide Resistance:	Chair:	Mark VanGessel, University of Delaware
		Russ Hahn, Cornell University
		Bill Curran, Penn State University
		Brian Olson, Dow AgroSciences
		Dan Kunkel, IR-4
		Dave Mayonado, Monsanto
Endowment Committee:	Chair:	Greg Armel, BASF
		Renee Keese, BASF
		Jeff Derr, Virginia Tech
		Erin Hitchner, Syngenta
		Hillary Sandler, University of Massachuesetts
		John O'Barr, BASF

2014 AWARDS, NEWSS ANNUAL MEETING

NEWSS Award of Merit: Russ Hahn, Cornell University. (Presented by Toni DiTommaso).

NEWSS Fellow Award: Renee Keese, BASF.

NEWSS Outstanding Researcher: Chuck Moehler, Cornell University

NEWSS Outstanding Educator: Dave Mortensen, Penn State University.

Robert D. Sweet Outstanding Graduate Student Award: Katherine Ghantous, University of Massachuesetts

M. Garry Schnappinger Service Recognition Award: Barb Scott, Univ. of Delaware

1st Place, Student Presentation: Matt Elmore, University of Tennessee.

2nd Place, Student Presentation: Clair Keene, Penn State.

2014 AWARDS, NEWSS ANNUAL MEETING (continued)

1st Place, Student Poster: Bryan Hicks, North Caolina State University (presented by Keith Burnell).

2nd Place, Student Poster: Zach Taylor, North Carolina State University.

2nd Place, Student Poster: Michael Cox, Virginia Tech.

1st Place, Photo Contest: Art Gover, Penn State (presented by Erin Hitchner).

2nd Place, Photo Contest: Bill Curran, Penn State.

3rd Place, Photo Contest: Jennifer Cote, University of Maine.

Transfer of the Presidential Gavel: incoming President Greg Armel, BASF (r), receives the gavel from outgoing President Dwight Lingenfelter, Penn State.

Presidential Service Recogntion: Past President Dwight Lingenfelter, Penn State, receives the honorary gavel from President Greg Armel, BASF.

2014 NEWSS COLLEGIATE WEED SCIENCE CONTEST

Hosted by Penn State University, July 29, 2014

1st Place Graduate Team: Virginia Tech (Team 1)
Sandeep Rana, Daniel Tekiela, Katelyn Venner

1st Place Graduate Individual:
John Brewer, Virginia Tech

2nd Place Graduate Team: NC State Univ. (Team 3)
Sam McGowen, Logan Grier, Shawn Beam

2nd Place Graduate Individual:
Kristy Perano, Cornell Univ.

3rd Place Graduate Team: Cornell University
Vinay Bhaskar, Courtney Stokes, Kristy Perano

3rd Place Graduate Individual:
Daniel Tekiela, Virginia Tech

2014 NEWSS COLLEGIATE WEED SCIENCE CONTEST (continued)

1st Place Undergrad Team: Univ. of Guelph (Team 2)
Chris Budd, Johanna Burrows, Tom Heeman, Stephen Boersma

1st Place Undergraduate Individual:
Stephen Boersma, Univ. of Guelph

2nd Place Undergrad Team: Univ. of Guelph (Team 1)
Matt Underwood, Josh Burrows, Jordan Eyamie, Jacob Nederend

2nd Place Undergraduate Individual:
Chris Budd, Univ. of Guelph

3rd Place Undergrad Team: Univ. of Guelph (Team 3)
Alex Vanhie, Andrea Smith, David Jefferson, Dylan Magnus with NEWSS President Greg Armel

3rd Place Undergraduate Individual:
Johanna Burrows, Univ. of Guelph

SEED BANK DYNAMICS OF INVASIVE SWALLOWWORTS. A. DiTommaso*, L.R. Milbrath, S.H. Morris, C.L. Mohler, and J. Biazzo, Cornell University, Ithaca, NY (1).

ABSTRACT

Pale swallowwort (SW) (*Vincetoxicum rossicum*) and black swallowwort (*V. nigrum*; Apocynaceae, subfamily Asclepiadoideae) are European viny milkweeds that have become invasive in many habitats in the northeastern U.S.A. and southeastern Canada. A multi-year seed bank study was initiated in fall 2011 to assess annual emergence and longevity of seed of pale SW and black SW at four different burial depths (0, 1, 5, and 10 cm) over four years. The experiment was conducted in pots buried with the rim nearly flush with the surrounding soil and filled with a locally collected soil. One hundred swallowwort seeds were sown in seed pans buried in each pot. Initial seed viability was 97% (black SW) and 95% (pale SW). Pots were checked twice weekly for swallowwort seedling emergence beginning in early May and weekly from July through September. Emerged seedlings were removed. Beginning in October 2012, seed pans were retrieved annually and recovered seed, seed coats and seed coat fragments were counted. Filled seeds (i.e., that appeared to have an embryo) were tested for viability with a 1% solution of tetrazolium chloride. The majority of seedling emergence occurred during the first year (92% in 2012), and no new seedlings emerged in the third year (2014). During the 2012 growing season, pale SW had relatively poor emergence at the 0 cm (12%), 5 cm (7%), and 10 cm (0.05%-only one seedling) sowing depths, while at 1 cm 38% of pale SW seeds emerged. The larger seeded black SW was more successful, with two-thirds of all sown seeds emerging at the 1 cm (71%) and 5 cm (66%) depths and 27% emerging at 10 cm. Only 17% of the surface-sown black SW emerged. A large portion of the seeds that germinated in the first year at 5- or 10-cm depths died before reaching the soil surface (pale SW-97%, black SW-48%). Of filled seeds that were recovered in 2012, mainly black SW at the 0 cm depth, 66% were viable. No viable seed was recovered after the second year. Key findings are that swallowwort seeds do not appear to survive more than two years in the soil, at least in our experiment, and that the two swallowwort species can germinate and emerge from soil depths of 10 cm which was not expected.

RESPONSE OF CORN AND SOYBEAN TO COMPETITION FROM IVYLEAF MORNINGGLORY UNDER WATER LIMITATION. R. Han and A. DiTommaso*, Cornell University, Ithaca, NY (2).

ABSTRACT

The impacts of climate change on water availability in crop production systems are becoming increasingly apparent. The increasing frequency and intensity of drought conditions can potentially alter competitive interactions between crops and weeds. Based on the physiology of the carbon fixation process, it is anticipated that C_4 plants might outperform their C_3 counterparts under water limiting conditions. A greenhouse study was conducted for 60 days to assess the competitive interactions between corn (*Zea mays*), a C_4 crop, and ivyleaf morningglory (*Ipomoea hederacea*), a C_3 vining annual weed, and between soybean (*Glycine max*), a C_3 crop and ivyleaf morningglory under an adequate supply and limiting water conditions. A single crop plant and 0, 1, 2, or 3 morningglory plants were grown from seed in 11.4 L pots filled with a standard Cornell soil mix. A single ivyleaf morningglory plant was also grown in pots in pure stand. The drought treatment consisted of suspending irrigation for five days after wilting was observed, providing water on the fifth day, and repeating the cycle three times for the corn trial and twice for the soybean trial. The amount of water provided on the fifth day of each cycle was determined by the quantity required to restore the soil water content of the zero weed pots to just above the critical wilting point. The corn and soybean trials were set up in a randomized complete block design with six replicates for each competition by water treatment combination. Aboveground biomass was reduced under the water limiting treatment for all three species relative to the adequate supply of water treatment, but the negative effect of water stress was more pronounced in ivyleaf morningglory than in the two crop species. Interspecific competition reduced the aboveground biomass of all three species relative to biomass produced in monoculture. In both corn and soybean, the presence, rather than density, of ivyleaf morningglory resulted in significant biomass loss. Ivyleaf morningglory biomass reductions were similar when grown with either corn or soybean. Our results suggest that corn and soybean growth may be more resilient under water limiting conditions than ivyleaf morningglory growth. Therefore, the negative impact of this weed on the yield of these two crops may be reduced, at least under our greenhouse growing conditions.

NARROW ROW SPACING IN WINTER WHEAT AS A TOOL FOR MANAGING ITALIAN RYEGRASS. Z.R. Taylor* and W.J. Everman, North Carolina State University, Raleigh (3).

ABSTRACT

Italian ryegrass (*Lolium multiflorum*) is one of the most problematic weeds in the production of winter wheat in the southeast. As herbicide resistance issues continue to develop and expand throughout the area, our options for control post emergence continue to decline. As a result we are looking at some cultural practices that may help to suppress the problem when combined with a variety of chemical control plans. One method studied was to improve the crops ability to compete by changing row spacing. We compared wheat planted in 7.5 rows with a drill to those planted in approximately 3.75 inch rows. Each row spacing received the following treatments; non-treated check, Zidua (pyroxasulfone) at 1.25 oz/a pre, Zidua at 1.25 oz/a pre fb Zidua at 1.25 oz/a post, Zidua at 1.25 oz/a pre fb Osprey (mesosulfuron) at 4.75 oz/a and non-ionic surfactant at 0.25% v/v post, Zidua at 1.25 oz/a pre fb Axial XL (pinoxaden) at 16.4 oz/a post, Zidua at 1.25 oz/a pre fb Osprey at 4.75 oz/a and Zidua at 1.25 oz/a and non-ionic surfactant at 0.25 % v/v post, Zidua at 1.25 oz/a pre fb Zidua at 1.25 oz/a and Axial XL at 16.4 oz/a post, Osprey at 4.75 oz/a and non-ionic surfactant at 0.25% v/v post, Axial XL at 16.4 oz/a post, Osprey at 4.75 oz/a and Zidua at 1.25 oz/a and non-ionic surfactant at 0.25 % v/v post, Zidua at 1.25 oz/a and Axial XL at 16.4 oz/a post, Axiom (flufenacet and metribuzin) at 8 oz/a at spike. Visual ratings did not show differences in control between the two planting arrangements. One location had very little ryegrass population, but population densities were recorded in the other location. Densities did not show differences between row spacing practices.

ROLE OF CORNELL COOPERATIVE EXTENSION IN INVASIVE SPECIES MANAGEMENT IN NEW YORK STATE: THE CORNELL COOPERATIVE EXTENSION INVASIVE SPECIES STATEWIDE OUTREACH PROGRAM. S. Bachman, C. Marschner*, D.G. Grantham, M.F. Wyman, E. Staychock, and C. O'Neill, Cornell University and Cornell Cooperative Extension, Ithaca, NY (4).

ABSTRACT

The mission of the Cornell Cooperative Extension Statewide Invasive Species Program (CCE ISP) is to provide New York stakeholders with research-based information on invasive species. We offer training and information support to New York's Partnerships for Invasive Species Management (PRISMs), county Cooperative Extension educators, and natural area managers. The goal of the CCE ISP is to improve New York's ability to prevent, detect and control new invasive species and manage existing ones. Key terrestrial species of focus include giant hogweed (*Heracleum mantegazzianum* Sommier & Levier), Japanese knotweed (*Polygonum cuspidatum* Siebold & Zucc.), phragmites [*Phragmites australis* (Cav.) Trin. ex Steud.], knapweed species (*Centaurea spp.*) and wild parsnip (*Pastinaca sativa* L.). In 2014, the CCE ISP is conducting a baseline survey of stakeholder knowledge of invasive species and motivation to protect our state from these invaders. This will provide vital information to further focus the CCE ISP's education, outreach and management plans.

As well as coordinating with our eight PRISM regions and 57 county Cooperative Extension offices, the CCE ISP cooperates closely with the CCE Invasive Species Clearinghouse (NYISC), the Invasive Species Research Institute (NYISRI), and the NY Invasive Species Database (iMAP Invasives). Each of these organizations plays a critical role in statewide invasive species control efforts. The PRISMs, based on the Cooperative Weed Management Area model, are regional invasive species management units, who prioritize invasive species threats to their region and coordinate control practices. County Cooperative Extension educators interface with the public and local organizations on invasive species. The Invasive Species Clearinghouse serves as a single, web-based location for information on invasive species, providing species profiles and information and links for the many organizations participating in invasive species control. The NYISRI coordinates research to help prevent and manage the impact of invasive species. They provide information on existing and new research, facilitate networking among scientists, and help standardize data collection protocols across research projects. The NY iMap partnership manages and shares invasive species data compiled from many different partners. Based on recommendations from the NY Invasive Species Task Force, the CCE ISP, PRISMs, NYISC, iMap and NYISRI were created to provide coordinated, effective, and efficient responses to invasive species challenges within the state.

CCE ISP collaborates closely with all of these partners. We share information and website resources with the Clearinghouse, draw upon the iMAP tools when providing outreach, and help gather invasive species data from researchers to expand the iMAP database. We collaborate closely with NYISRI by providing information on the knowledge needs of the PRISMs and CCE educators, and share NYISRIs research results with partners. Through these collaborations, the CCE ISP reaches out to all of New York State to fulfill the need for invasive species education and outreach for our diverse landscapes.

HIGH-RESIDUE CULTIVATION AS AN INTEGRATED WEED MANAGEMENT TACTIC IN NO-TILL CORN AND SOYBEAN. C.L. Keene* and W.S. Curran, The Pennsylvania State University, University Park (5).

ABSTRACT

Integrated weed management practices are needed to control herbicide-resistant weeds and must form the basis of robust weed control programs if they are to remain viable over the long-term. Shallow high-residue cultivation is a weed management tactic compatible with no-till corn and soybean production and provides a mechanical weed control option in no-till systems. An experiment was conducted at the Penn State Agricultural Research Center in Rock Springs, Pennsylvania to test the efficacy of high-residue cultivation timing and frequency on weed control in no-till corn (*Zea mays* L.) and soybean (*Glycine max* Merr.). The experiment was a randomized complete block split-plot design with cultivation timing and frequency constituting the main plot and the presence or absence of cover crop residue as the subplot. The cover crop preceding soybean was cereal rye (*Secale cereal* L.) and preceding corn was a mixture of hairy vetch (*Vicia villosa* Roth.) and triticale (*x Triticosecale* Wittm.). Cover crops were terminated with herbicides and rolled with a roller-crimper prior to cash crop planting. Cultivation treatments included an uncultivated weedy check, cultivated once at either 4, 5, or 6 weeks after planting (WAP), cultivated twice at 4 and 5, 4 and 6, or 5 and 6 WAP, and cultivated three times at 4, 5, and 6 WAP. An herbicide-only weed-free check was included in 2012 and 2013. The experiment was conducted in soybean in 2011, 2012, and 2013, and in corn in 2012.

In 2011 end-of-season weed biomass was similar in the 3-pass, all 2-pass, and the 1-pass at 5 WAP treatments: mean weed biomass for these treatments ranged from 13 to 335 kg ha^{-1}. Weed biomass in the once at 4 WAP was similar to the weedy check and weed biomass in the once at 6 WAP was lower than the weedy check but higher than the other cultivation treatments. Soybean yield in 2011 was similar to the highest yielding treatment (3.8 Mg ha^{-1}) in the 3-pass, all 2-pass, and the 1-pass at 5 WAP treatments and significantly lower in the once at 4 WAP and once at 6 WAP treatments (2.8 Mg ha^{-1}). Soybean yield was significantly influenced by cultivation treatment and residue but not by the interaction: yields were consistently higher in plots with rye residue than in plots with no residue. In 2011 maximum soybean yield was 4.1 Mg ha^{-1} and was achieved in the 2-pass at 4 and 5 WAP. In 2012 weed biomass was similar in all cultivation treatments and ranged from 20 to 507 kg ha^{-1} with the exception of the once at 4 WAP with residue which had significantly more weed biomass (910 kg ha^{-1}) than all other cultivation treatments. Soybean yield in 2012 was significantly influenced by cultivation treatment: yield was similar to the weed-free check (4.0 Mg ha^{-1}) for all cultivation treatments with the exception of the once at 6 WAP which was significantly lower (3.2 Mg ha^{-1}) and similar to the uncultivated weedy check (3.0 Mg ha^{-1}). In 2013 weed biomass in soybean was greater in plots without residue (850 kg ha^{-1} weed biomass) than with residue (433 kg ha^{-1}). Weed biomass was lowest in the 2-pass at 4 and 6 WAP (187 kg ha^{-1}) and similar to the other 2-pass and 3-pass treatments. Weed biomass in the single-pass treatments was similar across the three timings and ranged from 799 to 1058 kg ha^{-1}, once at 4 WAP and once at 5 WAP treatments, respectively. In 2013 soybean yield was similar across all cultivation treatments ranging from 3.2-3.7 Mg ha^{-1} and was only similar to the uncultivated weedy check (2.6 Mg ha^{-1})

at the once at 5 WAP treatment.

In corn in 2012, weed biomass in all cultivation treatments was similar to the lowest amount, 116 kg ha^{-1}, from the 2-pass at 5 and 6 WAP except the once at 4 WAP which was higher at 562 kg ha^{-1}. Corn yield was influenced by the presence of hairy vetch-triticale residue and was higher, 10.9 Mg ha^{-1}, in plots with hairy vetch residue than without, 10.1 Mg ha^{-1}. In 2012 corn yield was not influenced by cultivation treatment.

These results suggest that two passes with a high-residue cultivator can consistently achieve weed control equivalent to that of an herbicide-only program in no-till corn and soybean in Pennsylvania. While a single well-timed pass may achieve equivalent weed control to 2 passes, e.g. 2011 in soybean, this result was not consistent across years. The impact of cover crop residue on weed biomass depended on the year: in years with a dry summer (2011 and 2012), residue tended to increase weed biomass while in wet years (2013), residue tended to suppress weeds. This variability in the impact of residue was seen because biomass levels were less than 5000 kg ha^{-1}, a threshold needed for weed suppression by the rolled cover crop mulch.

WEED CONTROL WITH ALS HERBICIDE-RESISTANT GRAIN SORGHUM IN NORTH CAROLINA. L.J. Vincent*, W.J. Everman, Z.R. Taylor, A.M. Knight, T.E. Besancon, and A.M. Growe, North Carolina State University, Raleigh (7).

ABSTRACT

Creating a fit for grain sorghum in North Carolina cropping systems has proven a difficult challenge for several reasons, particularly in achieving effective weed management. However, DuPont Pioneer has introduced the Inzen-Z sorghum system to provide relief for sorghum growers. Inzen is an ALS-herbicide resistant grain sorghum variety created from traditional breeding methods and is paired with their proprietary herbicide, Zest. The proprietary herbicide Zest is a re-formulated liquid nicosulfuron product, the same active ingredient as Accent typically used for post emergence grass control in corn. The product is awaiting final approval from the Environmental Protection Agency (EPA) and the system as a whole is expected to be available for the 2015 growing season. Thus far, Zest has been said to control foxtail species, barnyardgrass, crabgrass, witchgrass, and non-ALS resistant shattercane. However, experimentation conducted this year expanded that weed control spectrum.

From summer 2014 field experimentation results; Zest has proven to provide excellent post emergence control of grass weeds in grain sorghum. In the 10 treatment protocol, four plots of each replication received a pre emergence application of s-metolachlor + atrazine. Of those four plots, Zest at a rate of 5 oz ai/A was combined with 2, 4-D + atrazine, pyrasulfotole + bromoxynil + atrazine, and dicamba + atrazine. The other five plots did not receive a pre emergence herbicide application and only received a MPOST herbicide application. Zest was applied at MPOST with atrazine, pyrasulfotole + bromoxynil + atrazine, clarity + atrazine, 2, 4-D + atrazine, and 2, 4-D + atrazine + metsulfuron. Regardless of timing, each plot received crop oil concentrate at 1% v/v and ammonium sulfate at a rate of 2 lb/A. The final plot of each replication was a non-treated check.

Based on herbicide efficacy ratings taken 7, 14, and 28 days after treatment (DAT) evaluating three different herbicide application timings; pre emergence (PRE), mid-post emergence (MPOST), and post emergence (POST); results were clear that including a pre emergence component is essential to staying weed-free. In every plot, texas panicum, yellow nutsedge, goosegrass, broadleaf signalgrass, and large crabgrass were better controlled in applications which contained a pre emergence herbicide application versus those which did not. At the alpha = 0.05 level, the PRE only herbicide application provided statistically significant better control of texas panicum and large crabgrass. In addition, the treatment tank mix of Zest + atrazine + pyrasulfotole + bromoxynil which received a PRE, proved to be a statistically significant leader in control of tough to tackle weeds such as texas panicum, goosegrass and compared to the same tank mix which did not receive a PRE. Finally, when comparing the Zest + 2,4-D + atrazine treatment which received a PRE and the treatment which did not, results are clear that including the PRE component provides a significant difference.

IMPACT OF ELEVATED CO_2 LEVELS ON CONTROL OF PURPLE AND YELLOW NUTSEDGE WITH GLYPHOSATE AND HALOSULFURON. C. Marble*, S.A. Prior, B. Runion, and A. Torbert, University of Florida, Apopka (8)

ABSTRACT

Atmospheric concentrations of carbon dioxide (CO_2) have been steadily rising each year and are expected to continue increasing in the future, which could have a significant impact on agricultural production. Previous research has shown that elevated CO_2 increases the growth and yield of most plant species. Although many crops may benefit from increasing levels of CO_2 in terms of growth and yield, the growth and distribution of many weed species is also likely to increase, as is the occurrence of herbicide resistant weed biotypes. There is now concern that global climate change could reduce the effectiveness of herbicides at current label rates. An experiment was conducted during the summer of 2012 to determine if labeled rates of glyphosate (RoundUp Pro®) and halosulfuron (SedgeHammer®) would provide effective control of purple nutsedge (*Cyperus rotundus* L.) and yellow nutsedge (*Cyperus esculentus* L.) in an enriched CO_2 environment. Tubers of yellow and purple nutsedge were grown in 3.0 L containers in either ambient or elevated (ambient + 200 $\mu mol\ mol^{-1}$) CO_2 in open-top chambers and treated with 0.5×, 1.0× or 1.5× the labeled rate of each product and a tank mix of each product at the specified rates. Both species of nutsedge had a positive growth response when exposed to elevated CO_2, as purple nutsedge increased shoot and root dry weights and tuber counts by 27%, 25%, and 25%, respectively, in comparison to plants grown at ambient levels. Yellow nutsedge had a greater response to elevated CO_2, increasing shoot, root, and tuber dry weights by 40%, 51%, and 100%, respectively, and tuber counts by 128%. Few differences were observed in terms of herbicide efficacy, and by 3 weeks after treatment provided over 90% control regardless of herbicide or tank-mix, rate, or CO_2 concentration. Based on the results of this study, it appears that although both weed species will increase growth and reproductive capability, the efficacy of halosulfuron or glyphosate will not be impacted by increasing atmospheric CO_2 levels. Both products provided adequate control of both nutsedge species when applied prior to the flowering stage. It is unclear how the efficacy of these active ingredients would have been impacted by delaying applications until either species was closer to or had reached maturity.

GLUFOSINATE AND GLYPHOSATE TOLERANT WEED MANAGEMENT SYSTEMS IN NORTH CAROLINA SOYBEANS. A.M. Knight*, W.J. Everman, and A. Simpson, North Carolina State University, Raleigh (9).

ABSTRACT

With increasing herbicide tolerance in the southeastern U.S., it is important for scientists to consider best management programs for herbicides on the market. Scientists must also consider the best way in which to preserve current modes of action for which resistance has not occurred. One way in which modes of action can be preserved is by utilizing crops with tolerance to herbicides of varying modes of action. Field studies were conducted in 2014 at the Upper Coastal Plain Research Station near Rocky Mount, NC comparing multiple glyphosate and glufosinate tolerant soybean varieties. These varieties were applied with herbicide programs of flumioxazin PRE, followed by a POST and late POST of S-metalochlor, fomesafen, and their respective tolerant herbicide. The study design was a randomized complete block. The success of these varieties with herbicide programs left minimal weeds for rating. Preliminary results indicated yield differences between treatments.

WEED COMMUNITY RESPONSE TO SOYBEAN SEEDING RATE AND COVER CROP MULCH RATE. C. Halde*, S. Mirsky, and M. Ryan, Cornell University, Ithaca, NY (10)

ABSTRACT

Cover crops can be used to overcome tradeoffs between soil health and weed management goals in organic cropping systems. Cover crops can facilitate reductions in tillage, and recent research has shown that no-till planting of organic soybean into rolled-crimped cereal rye can result in weed suppression and soybean yields that are equivalent to tillage-based management. As in conventional no-till systems, reductions in tillage in organic systems can result in weed community shifts toward perennial species. In this research, a range of cover crop mulch rates and soybean seeding rates were created to assess their weed suppressive ability and their effect on weed community composition and structure. The study was conducted in 2008 and 2009 in Maryland and Pennsylvania, using five levels of cereal rye residue representing 0, 0.5, 1, 1.5, and 2 times the ambient level, and five soybean seeding rates ranging from 0 to 74 seeds m^{-2}. A cereal rye cover crop was planted in the fall, then clipped and removed from the plots the following spring. Soybeans were seeded at five rates, before the clipped rye biomass was returned to the plots. Weed biomass decreased with increasing rye residue and decreased with increasing soybean density. Smooth pigweed (*Amaranthus hybridus* L.), wild buckwheat (*Polygonum convolvulus* L.), and four summer annual grasses were associated with the absence of soybeans. The effect of site-year was apparent in the NMDS ordination, indicating that weed community composition varied across site-years. Weed biomass ranged from 0 to 967 g m^{-2} across all site-years. Soybean yields were optimized with 100-1200 g m^{-2} of rye mulch and soybean seeding rates of 50-60 seeds m^{-2} in an organic no-till system. Anticipating and managing for weed community shifts is important for farmers, and even more critical in organic systems because of the reduced options for weed control.

IMPACT OF GROWTH REGULATOR RATE AND APPLICATION TIMING ON SORGHUM GROWTH AND YIELD. T.E. Besancon*, L.J. Vincent, and W.J. Everman, North Carolina State University, Raleigh (11)

ABSTRACT

North Carolina growers produced sorghum (*Sorghum bicolor* (L.) Moench) on 40,000 acres in year 2014. This acreage is expected to increase in the future due to an assured market, grower enthusiasm, low production cost, drought tolerance and a lowered risk when compared to corn, especially on sandy soils. Since sorghum is an old crop with renewed interest, this is the first time it has been grown on a large scale in North Carolina. This crop may be a good fit in the overall cropping system of the state. However, limited information on production and weed management is available in the state. Weed control is an important aspect of profitable crop production which has a bearing on the success of a new crop, and in determining its fit in a cropping system. In order to address the issues of successful weed management under the given environment and cropping system, studies were conducted from 2012 to 2014 at the Upper Coastal Plain Research Station (Rocky Mount, NC), Caswell Research Farm (Kinston, NC), and Central Crops Research Station (Clayton, NC). The objective was to evaluate the growth and yield response of sorghum to different rates of 2,4-D amine (100, 217 and 333 g $ai.ha^{-1}$) and one rate of dicamba (280 g $ai.ha^{-1}$) applied post-emergence beyond the recommended growth stage (15-20 cm tall), at 25, 35, 46, 56, 66, and 74 cm height. Growth regulator applications on 30 to 60 cm tall sorghum resulted in significant taller plants compared to later treatments. Consequently, an important lodging effect was observed when crop height reached 65 cm tall. Yield was impacted by growth regulator applications at locations where planting on sandy soils with a low field capacity resulted in reduced sorghum growth and increased crop sensitivity to herbicides and environmental stress. These results tend to confirm the higher crop sensitivity to growth regulators when sorghum is planted in soils with low water retention capacity. However, when planted in soils with higher water retention capacity, sorghum can tolerate growth regulators applied over-the-top up to 50 cm crop height at application without significant yield reduction.

DICAMBA-RESISTANT SOYBEAN WEED MANAGEMENT SYSTEMS. Anthony Growe, Dennis Williamson, Tony White, Wesley Everman, North Carolina State University, Raleigh (13).

ABSTRACT

Due to glyphosate-resistant weed biotypes becoming more common in North Carolina agricultural systems, new technologies, such as the development of dicamba-tolerant soybeans, will be available to integrate into current weed management systems. An experiment was conducted at the Upper Coastal Plain research station in Rocky Mount, NC to examine the effectiveness of weed management strategies involving experimental dicamba, dicamba premixes and flumioxazin in dicamba- tolerant soybeans. Various weed species, including Palmer amaranth (*Amaranthus palmeri*), were treated with various rates and herbicide combinations to determine greatest control. Plots were rated 23, 38 and 52 DAP. Results exhibited greater Palmer amaranth control when an effective PRE followed by a POST was utilized. All POST systems obtained statistically similar weed control. The experimental dicamba PRE followed by POST of dicamba/glyphosate premix had greater control than a PRE of dicamba alone. All treatments with a PRE followed by POST exhibited more than 89% control of Palmer amaranth.

RELATIVE CORN SAFETY WITH HIGH RATES OF PREEMERGENCE ALS-INHIBITING HERBICIDES. M.J. VanGessel*, B.A. Scott, and Q.R. Johnson, University of Delaware, Georgetown (14)

ABSTRACT

Soil-applied herbicides that contain acetolactate synthase inhibiting (ALS) (or Group 2) herbicides or isoxaflutole (Group 27) have not been compared for relative crop safety when used on coarse-textured soils. Furthermore, most weed control trials are not designed to evaluate the impact of herbicide injury on corn yield. This trial compared relative crop safety of soil-applied herbicides containing ALS-inhibiting herbicides or isoxaflutole, and the resulting impact on yield.

The trial was conducted in 2013 and 2014. The study was a strip plot with herbicides as the whole plot and corn hybrid as the sub-plot. Herbicide treatments were preemergence applications of Hornet WDG Premix, Basis Blend Premix, Capreno Premix, Prequel Premix, Corvus Premix, Balance Flexx, Instigate Premix, Instigate Premix followed by postemergence (POST) application of Realm Q Premix, Realm Q Premix applied POST, and an untreated plot. Herbicides were applied at twice the labeled rate. Plots were maintained weed free.

Field corn (*Zea mays* L.) was planted May 6, 2013 and April 23, 2014 and herbicides applied within 24 hours of planting. Hubner hybrids 'H4744', 'H5555' and 'H6179' were used. Hybrids were selected to provide a range of sensitivity to Groups 2 and 27 herbicides, with H5555, listed in the company's literature as sensitive to Group 2 and 27.

In both years, the site received over 1.2 inch of rainfall within one week of planting. Irrigation was used to avoid mid to late season moisture stress.

Corn injury was significant for the main effects of hybrid and herbicide. Injury was rated weekly until 9 weeks after planting. The hybrid's recovery from herbicide injury was determined by calculating the area under the curve. While the ranking of treatments differed across years, Balance Flexx had the least amount of injury, with no more 10% at any rating. The recover value for Balance was also the lowest in both years, while Basis, Prequel, Instigate, and Hornet had the highest recovery values. Higher values indicate slower recover rates.

The corn hybrid, H5555, tended to exhibit more injury than the other two hybrids in the trial. This hybrid is rated by the seed company as sensitive to Group 2 herbicides.

Yield for all herbicide treatments, based on bu/A, was not significantly different from the untreated check. Under unfavorable conditions, (coarse textured soils, overlaps, lower temperatures, and high rainfall prior to corn emergence) corn injury from some soil-applied herbicides can be severe. However, yield loss may not be as severe as early-season injury indicates if environmental conditions during mid and late season are favorable.

CORN WEED MANAGEMENT PROGRAMS FOR TROUBLESOME WEEDS IN NORTH CAROLINA. B.W. Schrage* and W.J. Everman, North Carolina State University, Raleigh (15)

ABSTRACT

The increase in corn acreage in North Carolina has proven to be a beneficial rotational crop in production systems determined to limit the presence of troublesome and herbicide-resistant weed species. To explore methods of optimizing control strategies, a study was conducted in Rocky Mount and Kinston, NC in 2014 evaluating certain residual overlapping residuals in maize. 69 DAT, *amaranthus palmeri* control was similar among all treatments ranging from 80-98%. Additional weed species were similarly managed by all treatments except for cases which experienced lesser control when applied with treatment 4 [atrazine + pyroxasulfone (Anthem ATZ) fb glyphosate (Roundup Powermax)] or 6 [thiencarbazone-methyl (Corvus) + atrazine (Aatrex 4L) fb tembotrione (Laudis) + atrazine + glyphosate]. In Rocky Mount, NC treatments 4 and 6 provided 21-47% less control of ivyleaf morningglory (*Ipomoea hederacea)*. There was no evidence of yield differences; however, treatment 4 resulted in 16-20% less control of pitted morningglory (*Ipomoea lacunosa)* 37 DAP in Kinston, NC. Sublevel control of hairy crabgrass (*Digitaria sanguinalis)* was also achieved by treatments 4 and 6 72 DAP.

SAFETY OF PRE-MIXTURES OF GLYPHOSATE PLUS RESIDUAL HERBICIDES TO LANDSCAPE PLANTS. J.C. Neal* and C.D. Harlow, North Carolina State University, Raleigh (16)

ABSTRACT

"Home owner" formulated combination POST with residual herbicides were compared for safety to semi-dormant herbaceous perennials and as directed applications around woody landscape plants. Treatments included Roundup Extended Control concentrate (glyphosate + imazapic) at 20 and 40 oz /A; Durazone concentrate (glyphosate + diquat + indaziflam) at 16 and 32 oz/A, and Spectracide weed and grass control at 17 and 34 oz/A. These treatments were compared to Roundup Quick-Pro (glyphosate + diquat) at 1 lb ai/A of glyphosate; applied alone or tank mixed with Barricade 4L (prodiamine) at 0.75 lb ai/A, as well as a non-treated control.

The first experiment was conducted on a planting bed that had been mulched the previous year with composted leaves. Container grown daylily and lirope plants were planted on February 9 and 10; 3 daylily and 6 liriope per plot. Daylily plants were dormant or had only a very small amount of new growth at the time of treatment. Lirope were dormant, but still had the remnants of last year's leaves on the plants which was trimmed to 3 inches. Treatments were applied on February 23rd. Before treatment half of the liriope plants in each plot were covered with a waxed paper cup to prevent foliar exposure. Cups were removed after treatment. Daylily plants, which were essentially dormant at the time of treatment, exhibited little or no injury. Liriope plants that received foliar applications of herbicides exhibited some initial injury. However, even plants covered at the time of treatment exhibited some injury from Durazone, Spectracide and Roundup Extended Control. No such injury was observed from Roundup QuickPro, suggesting that the injury to covered lirope resulted from root uptake of one of the active ingredients in these consumer products. Roundup Extended Control caused over 80% injury to both covered and uncovered liriope with of covered plants. Durazone caused up to 85% injury to un-covered liriope but less than 20% injury to uncovered plants.

A second experiment was established on woody landscape plants. Loropetalum (*Loropetalum chinensis* 'Ruby') and flowering dogwood (*Cornus florida* 'Cherokee Chief') were planted into a clay loam soil amended with rotted leaves on late November 2012 then mulched with composted wood chips. On May 3 and October 15 2013 treatments were applied as a directed spray avoiding contact with plant foliage. Treatments were the same as in the prior study except Spectracide was not included. The experimental design was as a randomized complete block with 3 plants per plot; each species planted in separate beds. Plants were observed monthly for injury and in the fall of 2013 loropetalum height and width were measured. From May through October no injury was observed on dogwoods. However, on November 14th plants treated with Roundup Extended had noticeably greater leaf drop. A visual estimate of % defoliation revealed that Durazone-treated, Quick-Pro treated and non-treated plants averaged about 20% defoliation; whereas, plants treated with Roundup-Extended averaged over 60% defoliation. Loropetalum was stunted by Roundup Extended Control applications but not by other treatments. The stunting was slow to develop and was not obvious to a casual observer until mid-summer. By October loropetalum plants in Roundup-Extended plots were 75% smaller than non-treated plants. Symptoms of injury included stunted growth and witches brooms. Treatments were reapplied in October 2013 but evaluations could not continue in the spring due to a late frost, which killed some dogwoods and damaged many of the loropetalum.

CONTROLLING JAPANESE STILTGRASS IN LAWN TURF. J.R. Brewer*, S.D. Askew, and S.S. Rana, Virginia Tech, Blacksburg (17)

ABSTRACT

Japanese stiltgrass (*Microstegium vimineum*) (JSG) is a nonnative, grassy weed that has become a severe problem in the Eastern United States. Native to Asia, JSG was introduced to the US in 1919 and can thrive in various types of environments from New York to Florida, including wetlands, woodlands, lawns, landscape beds, and mountainous regions. Due to its flexible habitat tolerance and ability to grow under both shade and full sunlight, JSG is threatening native understory species. Research has been conducted to determine the most efficient and economic chemical or mechanical methods to control JSG. Studies in the past have used selective herbicides and mechanical controls, including fenoxaprop, imazapic, sethoxydim, mowing, hand pulling, and others to evaluate JSG control in natural areas such as wetlands and forest but not in a lawn or turf setting. Our study in Newport, Virginia was conducted on a residential lawn to determine the most effective chemical means to control JSG without injuring the desired turfgrass. The trial was initiated on August 15, 2014 on a mixed Kentucky bluegrass (*Poa pratensis*) and tall fescue (*Lolium arundinaceum*) lawn. Treatments included: mesotrione at 280 g ai/ha^{-1} once, mesotrione at 140 g ai/ha^{-1} twice, topramezone at 57 g ai/ha^{-1} once, topramezone 24 g ai/ha^{-1} twice, quinclorac at 1120 g ai/ha^{-1} once, quinclorac at 660 g ai/ha^{-1} twice, fenoxaprop at 140 g ai/ha^{-1} once, fenoxaprop at 70 g ai/ha^{-1} once, fenoxaprop at 35 g ai/ha^{-1} once, triclopyr at 1120 g ai/ha^{-1} once, mesotrione at 280 g ai/ha^{-1} + triclopyr once, topramezone at 54 g ai/ha^{-1} + triclopyr once, and all treatments included an adjuvant(NIS or MSO) except fenoxaprop and triclopyr. These treatments were applied with a hooded sprayer at 45 L/ha^{-1} and 4.8 km/h, and the sprayer had a 71.12 cm spray width.

Initial JSG cover in the lawn plots ranged from 30 to 80%. At 3 WAIT, mesotrione, quinclorac, and triclopyr controlled JSG less than 40%. Topramezone at 24 g ha^{-1}, combinations of topramezone or mesotrione with triclopyr, and fenoxaprop at 14 or 28 g ha^{-1} controlled JSG at least 67%. AT 6 WAIT, JSG had recovered from most treatments. Fenoxaprop at any rate completely controlled JSG. All topramezone-containing treatments controlled JSG >70%, and higher than the remaining treatments.

COMPARISON OF TWO WEEDING STRATEGIES IN CONTAINER NURSERIES. C.D. Harlow* and J.C. Neal, North Carolina State University, Raleigh (18)

ABSTRACT

Weed control is a costly component of nursery crop production, and recent economic pressures have forced growers to search for ways to reduce production costs. Despite the availability of many effective preemergence herbicides, weeds continue to emerge in container nursery crops, necessitating costly removal by hand. Anecdotal evidence suggests that increasing the frequency of hand weeding prevents weeds from going to seed and can in the long run reduce overall weed populations and total hand weeding time. This theory was tested in two experiments, one at a cooperating container nursery and one at a research station. Methods were similar at each site. Experimental treatments were two different hand weeding strategies – removing weeds every 2 weeks or removing weeds every 8 weeks. At the research station each experimental unit included about 1000 4-L pots, and at the on-farm site each experimental unit consisted of a production block of over 5000 10-L pots. Plants were potted in late spring, and all pots were treated with Snapshot TG at 150 lb/A at the initiation of the experiment and every 8 weeks thereafter. At each 2-week weeding, the weeding crew was instructed to remove only those specimens which had grown large enough to be removed easily or appeared to have the potential to flower. At the 8-week weeding, all weeds were removed and the surface of the substrate "raked" by hand to dislodge small seedlings. Man-hours required for weeding and fresh weights of weeds removed were recorded. The following results are from the first two 8-week cycles; experiments will continue until plants are marketable. Some differences between sites were observed. Weed pressure was more variable at the grower site than at the research station site where plots were over-seeded with a mixture of common weeds to promote uniformity. The fresh weights of weeds removed from pots were significantly less in plots weeded every 2 weeks. Cumulative weed weights at the research station were 0.7 Kg for plots weeded every 2 weeks and 6.3 Kg for plots weeded every 8 weeks. Weed weights at the on-farm site were 6.7 Kg for plots weeded every 2 weeks and 13.7 Kg for plots weeded every 8 weeks. Following the first 8 weeks in the research station experiment, cumulative time (man-hours) to remove weeds in the 2-week treatment was about half of that required in the 8-week treatment. The difference was not as pronounced following the next 8 weeks. The overall trends supported the hypothesis that increased weeding frequency reduced weed populations and biomass, but whether this translates to an economic return by significantly reducing hand weeding labor costs is yet unclear.

KENTUCKY BLUEGRASS CULTIVAR RESPONSE TO METHIOZOLIN. S.S. Rana*, S.D. Askew, and S. Koo, Virginia Polytechnic Institute and State University, Blacksburg (19)

ABSTRACT

Methiozolin is a new isoxazoline herbicide under development by Moghu Research Center for use in turfgrass in the US and other countries. Methiozolin is typically used for selective POST annual bluegrass (*Poa annua*) control on creeping bentgrass (*Agrostis stolonifera*) putting greens. However, previous research at Virginia Tech and Moghu Research Center suggests that methiozolin at higher rates (1000 g ai/ha applied twice at monthly intervals) can also be used for annual bluegrass control on fairways. Controlling annual bluegrass on Kentucky bluegrass (*Poa pratensis*) fairways has always been challenging due to physiological similarities in the genus *Poa*. Moreover, Kentucky bluegrass is known for variation in herbicide response between cultivars. Since most previous work has been focused on the use of methiozolin on putting greens or green surrounds, more research is needed to elucidate methiozolin use on fairways. Therefore, field trials were conducted from fall 2011 through spring 2012 and from fall 2013 through spring 2014 at the Virginia Tech Turfgrass Research Center (TRC) in Blacksburg, VA, to determine response of 110 Kentucky bluegrass cultivars to methiozolin. Permission was granted to use a five-year-old Kentucky bluegrass National Turf Evaluation Program trial site for this study. The site was maintained at a 1.5-cm mowing height and fertility, pest control, and other inputs were added as needed to simulate golf fairway maintenance. Each 1.5 m by 1.5 m Kentucky bluegrass cultivar plot was subdivided into 4 subplots, which were treated twice at a one-month interval with methiozolin at 0 (NTC), 2000 (2X), 4000 (4X), and 6000 (6X) g ai/ha, respectively. The site lay dormant for one year between trial repetition and subplots were rotated 90 degrees and re-randomized for the second trial. Kentucky bluegrass, annual bluegrass, and broadleaf weed cover/control ratings were assessed 3 times each in fall and spring following methiozolin applications. Response of Kentucky bluegrass cultivars and weeds to methiozolin was similar in both years; therefore, data were pooled across years. At 6 weeks after initial treatment (WAIT), methiozolin at the 2 and 4X rates was safe to all 110 Kentucky bluegrass cultivars. The 6X rate of methiozolin injured approximately 15% of the 110 Kentucky bluegrass cultivars more than commercially acceptable levels (30%) and injury was expressed as stunting with little or no discoloration to turf. At 21 WAIT, methiozolin at all rates tested produced Kentucky bluegrass cover equivalent or better than the non-treated check. All rates of methiozolin reduced annual bluegrass, corn speedwell (*Veronica arvensis*), Persian speedwell (*Veronica persica*), common chickweed (*Stellaria media*), and mouse-ear chickweed (*Cerastium vulgatum*) cover 21 WAIT. Results from this study indicate that methiozolin controls annual bluegrass and several other broadleaf weeds with commercially-acceptable Kentucky bluegrass response on all cultivars at 2 times and most cultivars at 4 and 6 times the rate needed for annual bluegrass control.

POISON IVY MANAGEMENT IN CRANBERRY. K.M. Ghantous* and H. Sandler, University of Massachusetts Amherst, Amherst (20)

ABSTRACT

Poison ivy (PI; *Toxicodendron radicans*) is a woody perennial weed that it is difficult to control in cranberry production areas, spreads quickly and can cause significant loss of crop. In addition to crop losses associated with PI, it also poses a health risk to farm workers who may be allergic to the plant. Glyphosate wipes are not feasible due to the lack of height differential between the weed and the crop canopy, and no other currently available herbicide provides satisfactory control.

Although tank mixing herbicides is a common practice in other crops, it is rarely done in cranberry. Mesotrione used at broadcast rates (600 ml ha^{-1}) has not been observed to affect PI, but anecdotal grower observations indicated that repeated spot treatment applications of a tank mix of sethoxydim and mesotrione will control poison ivy populations. A mixture of 11.7 ml of each herbicide and 11.7 ml of crop oil concentrate (COC) per L was applied at 4 different cranberry farms infested with poison ivy. Five treatment scenarios were tested: two "early" applications (late May and mid-June) in 2013 only, two "early" applications in 2013 and again 2014, two "late" applications (mid-June and early July) in 2013 only, two "late" applications in 2013 and again 2014, and untreated. Percent PI and cranberry cover were measured using point quadrats, and cranberry yield was collected.

In September 2014, PI cover was significantly lower in plots treated late 2013, late both years and early both years compared to the untreated but treatments did not differ from one another; cranberry cover was also significantly greater for these treatments. The effect of treatments on yield varied from site to site. This could be due to difference in cranberry variety, as well as differences in crop health prior to treatment. Cranberry phytotoxicity was observed in 2013, possibly due to injury from COC use in high air temperatures. In general, 2013 yields from early and late treatments resulted in a decrease in fruit compared to the untreated areas. Cranberry phytotoxicity was not observed in 2014 (treatments applied during acceptable temperature conditions). For 2014 yields in general, plots treated in 2013 only and in both years had higher yields than untreated plots, but the effect of treatment timing varied by site.

After achieving significant PI control in 2013 with the tank mix, an additional field experiment was conducted in 2014 to evaluate the efficacy of COC or nonionic surfactant (NIS) alone, sethoxydim + COC, and mesotrione with either adjuvant, and the herbicide combination with either adjuvant for PI control. The same herbicide rates were used as in the original study, and treatments were applied at the same timing as the "early" treatments. Adjuvants alone and sethoxydim + COC were all ineffective for reducing PI cover. Treatments of mesotrione with either adjuvant, as well as mesotrione in combination with sethoxydim and either adjuvant, were equally effective at reducing PI cover. There were no significant yield differences. Preliminary results indicate that mesotrione can be used with either COC or NIS as a spot treatment to reduce PI cover. Additional research is needed to determine if the addition of sethoxydim improves PI control, and also to determine if lower rates of mesotrione will provide adequate control.

EVALUATION OF PREEMERGENCE HERBICIDES FOR SEVERAL CONTAINER GROWN ORNAMENTAL GRASSES. A. Senesac*, J.C. Neal, and J.F. Derr, Cornell Cooperative Extension, Riverhead, NY (21)

ABSTRACT

Container studies were conducted at three research stations: Cornell University, Long Island (Riverhead, NY), North Carolina State University (Castle Hayne, NC) and Virginia Tech (Hampton Roads, Virginia Beach, VA) to evaluate the response of commonly grown ornamental grass and sedge species to preemergent herbicides.

At the Riverhead site, two common pre-emergence herbicides: pendimethalin and dithiopyr were evaluated on 11 ornamental grass and sedge species. The study was established in early June, 2014 when the well established seedling plugs were transplanted shortly before treatment. Ornamental grass and sedge species were evaluated for tolerance to two herbicides. Pendimethalin only: *Ammophila breviligulata, Chasmanthium latifolium* and *Muhlenbergia capillaris*; dithiopyr only: *Deschampsia cespitosa* and *Festuca glauca* 'Elijah Blue'; or pendimethalin and dithiopyr: *Carex buchanani* 'Red Rooster', *Carex morrowii 'Ice Dance', Hakonechloa macra* 'Aureola', *Leymus arenarius* 'Blue Dune', *Pennisetum alopecuroides, Pennisetum orientale, Schizachyrium scoparium* and *Sporobolus heterolepis.* The rates of each herbicide were: 1X, 2X and 4X labeled use rate. Pendimethalin (2G) was applied at 3, 6 and 12 lbs./A (a.i) and dithiopyr (2EW) at 0.5, 1.0, and 2.0 lbs./A (a.i.). The treatments were applied on June 13th and again six weeks later. The results suggest that most of the species were quite tolerant of the herbicide treatments. The species that were injured only showed it after the second treatment and at the higher application rates. Dithiopyr caused injury to *Pennisetum orientale, Schizachyrium scoparium, Sporobolus heterolepis* and *Festuca glauca* 'Elijah Blue'.

At the North Carolina site, the grasses in the experiment included: little bluestem (*Schizachyrium scoparium*), northern sea oats (*Chasmanthium latifolium)*, Calamagrostis (*Calamagrostis x acutiflora* 'Karl Foerster'), switchgrass *(Panicum virgatum)*, and pink muhly grass (*Muhlenbergia capillaris*). Treatments included isoxaben (Gallery SC) at 1, 2, and 4 lb. a.i./A, Pendulum 2G at 3, 6, and 12 lb. a.i./A, and dithiopyr (Dimension EW) at 0.5, 1, and 2 lb. a.i./A). Additionally combination treatments of isoxaben (1 lb. a.i./A) with dithiopyr (0.5 lbs./A) or pendimethalin (3 lbs./A) were included. Plants were cut back to about five inches six weeks after treatment and treatments were re-applied. No acute injury was observed from any treatment. However, over time several species exhibited some reduced growth from some treatments. *Calamagrostis* growth was slightly reduced by Dimension at 2X and 4X rates but plants recovered. Little bluestem growth was slightly reduced by the 4X dose of Dimension EW and rooting of the tillers was reduced by the 2X and 4X doses, as evidenced by the number of plants with a “wobbly top”. Twelve lb. a.i/A of Pendulum 2G also reduced tiller rooting. *Panicum virgatum*, *Muhlenbergia* and *Chasmanthium* were uninjured by these treatments. In nursery situations it is likely that combinations of Gallery plus a “grass” herbicide such as Dimension or Pendulum would be utilized. Combination treatments applied at the labeled doses did not reduce growth or quality of any species tested.

At Hampton Roads, VA the species evaluated were: C*hasmanthium latifolium*, *Pennisetum alopecuroides* (*Calamagrostis x acutiflora* 'Karl Foerster'), *Schizachyrium* *scoparium*, *Panicum virgatum*, and *Eragrostis curvula*. At 36 days after treatment, no herbicide treatment reduced shoot fresh weight of *Schizachyrium, Panicum, Eragrostis, Pennisetum,* or *Chasmanthium.* Dimension at 2 lb. a.i/A reduced shoot weight of *Calamagrostis* compared to untreated plants, and the two lower rates of Dimension appeared to reduce shoot weight in this species. After the second herbicide application, no treatment reduced regrowth shoot weight of *Schizachyrium, Panicum, Eragrostis,* or *Pennisetum* although Pendulum-containing treatments appeared to cause approximately a 10% reduction in *Schizachyrium* regrowth weight. Dimension at 2 lb. a.i/A appeared to reduce regrowth shoot weight of *Calamagrostis*. When compared to untreated plants, no herbicide treatment reduced seedheads production in *Schizachyrium, Panicum, Pennisetum,* or *Chasmanthium*. *Eragrostis* and *Calamagrostis* did not produce seedheads during the trial. *Pennisetum alopecuroides, Calamagrostis acutiflora, Schizachyrium scoparium, Panicum virgatum*, and *Eragrostis curvula* tolerate the herbicides used in this study. *Chasmanthium latifolium* tolerated all herbicide treatments after one application, but there appeared to be about a 20% reduction of growth after two applications for the herbicides tested.

FIRST-YEAR RESULTS OF A SYSTEMS COMPARISON OF CULTIVATION-BASED VERSUS MULCH-BASED WEED CONTROL IN YELLOW ONION. B. Brown* and E. Gallandt, University of Maine, Orono (22)

ABSTRACT

Successful weed management may be achieved by 1) intensive, seedling-focused management involving repeated cultivation during the "critical weed free period" of the crop; 2) comprehensive seed-focused management with a goal of zero seed rain; or weed prevention through mulch, either black plastic mulch (with 3) cultivated paths or 4) straw mulched paths) or organic mulch of 5) straw or 6) hay. While each of these approaches may be able to maintain crop yields, they vary in their short- and long-term economics as well as the relative balance of soil aggrading versus degrading practices employed. To evaluate the whole system effects of the strategies, each was used as a treatment during summer of 2014 using yellow storage onions (*Allium cepa* cv. 'Cortland') as a test crop while labor, soil quality, weed suppression, and yields were recorded. We found that weed growth had the strongest effect on yield ($P<0.001$). The critical weed free period treatment had the most mid- and late-season weeds and yielded significantly less than the other strategies (39,200 lbs/acre compared to 58,400 lbs/acre), contrary to literature suggestions. Weeding labor was greatest in the zero seed rain treatment (171 hours/acre) due to the extensive amount of cultivation needed to keep weeds from setting seed. The organic mulch treatments required the most total labor (311 hours/acre) due to the time required to hand-apply the mulch. We also found several differences in soil quality among treatments; the mulch treatments exhibited significantly less late-season compaction as well as increased water infiltration. Afternoon soil temperatures were an average of 10.6 degrees F cooler in the organic mulch than the plastic mulch and zero seed rain treatments. The late-season critical weed free period treatment had similar soil temperature as the organic mulches due to the shade of the weeds.

PERFORMANCE AND MANAGEMENT OF INTER-SEEDED COVER CROPS IN A TROPICAL SEMI-ARID CROPPING SYSTEM AND IMPACT ON YIELD AND WEEDS.
V. Bhaskar*, R. Bellinder, and M. Walter, Cornell University, Ithaca, NY (23)

ABSTRACT

This study was conducted to investigate the potential for inter-seeded cover crops in a tropical semi-arid monocropping system in the Vidarbha region of India. Here, the following factors have caused soil organic carbon reserves to fall to unsustainable levels: (1) semi-intensive farming involving copious tillage, both before planting to break up the hard clayey soils and after planting (May-June) for weed management, (2) removal of crop residue from fields following harvest in January and (3) summers (February to May) dry and hot enough that farming is seldom done and even weeds don't come up, thus leaving the soil barren and exposed. Cover crops would help to improve the soil conditions, however, in an intercropped system, the strategies adopted to manage them and minimize their effect on the cash crop influences the outcome of an introduction into an existing cropping system.

Field trials were carried out in monocropped cotton (*Gossypium hirsutum*) with five cover crop species as the treatments: sesbania (*Sesbania spp.*), gliricidia (*Gliricidia sepium*), a sorghum sudan grass hybrid, lablab (*Lablab purpureus*) and sunnhemp (*Crotolaria juncea*). The objectives of the experiment were to understand (1) the effect on cotton measured in terms of plant height, number of bolls and most importantly, yield, (2) weed suppression measured in terms of weed ground cover and density, and (3) cover crop performance measured in terms of biomass production, ground cover and stand density.

The most important cover crop management practice was to cut back the cover crop to a height suitable for their regeneration every time they started to shade the cotton canopy. Cotton being very slow growing initially, taking about 3 to 4 months to achieve total ground cover, provides excellent opportunity for the cover crops to establish well enough to withstand later shading and part of the summer after harvest of the cotton crop. The field trials were carried out for two years at 4 different sites including 3 on farm and one at the local agricultural research station. The results showed that the treatments did not have a significant effect on cotton yield, suppressed weeds adequately in a good stand and produced dry biomass in amounts of 5 to 12 tons/ha. It was observed that when cover crops received 1 to 3 weed-free weeks after sowing and exhibited a vigorous stand, suppression of weeds was good, with densities less than $15/m^2$ without any form of mechanical or chemical inter row weed control. It was further noticed and later confirmed upon analysis of data, that good cover crop stands led to a slight decrease in the number of bolls but with a corresponding increase in boll size as well as a slight increase in cotton plant height. Both these changes in boll size and plant height can be considered advantageous. Sunnhemp and sesbania were the more promising species with quick emergence, vigorous initial growth, non- sporadic stands, rapid ground cover, adequate weed suppression and good biomass production.

CONTROL OF ANNUAL GRASSES IN TURF USING THE BIOHERBICIDE THAXTOMIN A (MBI-005). J.C. Wolfe*, J.C. Neal, and C.D. Harlow, North Carolina State University, Raleigh (24)

ABSTRACT

Increasing consumer and regulatory pressure in recent years have led to increasing interest in alternatives to synthetic chemical pesticides in both the United States and abroad. However, there are currently very few non-synthetic products capable of providing selective weed control in turf. Thaxtomin A (MBI-005) is a metabolite of the bacterium *Streptomyces scabies,* which has been identified as a candidate for commercialization as a bioherbicide. Past research has shown that thaxtomin A functions as a cellulose biosynthesis inhibitor, providing selective preemergence and postemergence control of broadleaf weeds without damaging established turfgrass. More recent studies, however, have demonstrated that thaxtomin A can also inhibit the germination of newly-seeded grasses, suggesting it may also have the ability to control problematic annual grasses such as annual bluegrass (*Poa annua*) and smooth crabgrass (*Digitaria ischaemum*).

The objectives of these studies were to determine if applications of thaxtomin A could provide equivalent control of annual bluegrass and smooth crabgrass to that of a synthetic preemergence herbicide. Experiments were conducted on a low-maintenance fairway at Thorndale Country Club in Oxford, NC in 2013 and 2014, and were initiated on March 28, 2013 and April 4, 2014 on smooth crabgrass and on September 18, 2013 and September 17, 2014 on annual bluegrass. All experiments were arranged in a randomized complete block design, with four replicates, and were evaluated monthly for weed control efficacy and percent weed cover. In 2013, treatments on both smooth crabgrass and annual bluegrass included thaxtomin A at 190 and 380 g ha^{-1} applied once, twice, and three times at four week intervals, along with an initial application of 380 g ha^{-1} followed by one and two applications at 190 g ha^{-1}. These treatments were compared to an industry standard preemergence herbicide, pendimethalin, applied once for annual bluegrass and twice at 8 week intervals for smooth crabgrass at 1.68 kg ai ha^{-1}. Treatments were also compared to corn gluten meal, a commercially available bioherbicide, applied twice at 8 week intervals at a rate of 980 kg ha^{-1}. A non-treated check was also included. In 2014 treatments were added to the smooth crabgrass experiment, and included four applications of thaxtomin A applied every four weeks at 190 and 380 g ha^{-1}, along with a treatment consisting of an initial application of 380 g ha^{-1} followed by three applications at 190 g ha^{-1} at four week intervals. Corn gluten was not included as a treatment in the 2014 experiments, as it did not control smooth crabgrass or annual bluegrass in 2013.

In 2013, three applications of thaxtomin A at 380 g ha^{-1} provided control of smooth crabgrass equivalent to that of pendimethalin through July, but by August control in these plots had declined. In 2014, four applications of thaxtomin A at 380 g ha^{-1} provided similar control through the month of August, but declined in September.

Three applications of thaxtomin A at 380 g ha^{-1} provided control of annual bluegrass equivalent to that of pendimethalin through May in the 2013 study, as did a single initial application at 380 g ha^{-1} followed by two applications at 190 g ha^{-1}. This study is currently being repeated.

These results suggest that thaxtomin A can be effective in providing commercially acceptable annual bluegrass control in turf. Acceptable control of smooth crabgrass can also be maintained through much of the season, but cannot be maintained throughout the entire growing season with four applications. Further study into the potential for season-long smooth crabgrass control with five or more applications is needed.

EFFECT OF PLANT AGE ON WEED CONTROL WITH MICROWAVE RADIATIONS. A. Rana* and J. Derr, Virginia Tech, Virginia Beach (26)

ABSTRACT

There is a demand for alternatives to chemical control of weeds, especially among those interested in organic approaches. The use of microwave radiations as a weed control method appears to be a good alternative because they do not produce chemical residues in the environment. Research was conducted to determine the impact of plant age on weed control using microwave radiations. Ten weed species representing monocots and dicots were selected for this study: southern crabgrass (*Digitaria ciliaris (Retz.) Koel*), dallisgrass (*Paspalum dilatatum Poir.*), yellow nutsedge (*Cyperus esculentus L,*), fragrant flatsedge (*Cyperus odoratus L.*), false green kyllinga (*Kyllinga gracillima Miq.)* , henbit (*Lamium amplexicaule L.)*, field bindweed (*Convolvulus arvensis L.),* common ragweed (*Ambrosia artemisiifolia L.*), white clover (*Trifolium repens L.*), and pitted morningglory (*Ipomoea lacunose L.*) were used in this study. The perennial species tended to show greater tolerance than the annual weed species tested. At the higher application rate, the broadleaf weed species tended to be more susceptible than the monocot species. Of the species tested, pitted morningglory and common ragweed were the most sensitive to microwave radiations, with complete control observed when plants 4 to 6 weeks old were treated at the lower application rate. Most of the monocot species regrew when treated with 90 or 180 joules/cm^2 of microwave radiations. In general, weed species became more tolerant of microwave radiations treatment as they increased in size. Averaged across species, the 8 to 10 week old plants showed more tolerance to microwave radiations treatment than plants 4 to 6 weeks old. The increase in biomass over time probably increases the amount of microwave radiations necessary for heating samples to the thermal threshold required for injury.

EFFECT OF WEED MANAGEMENT ON NITROUS OXIDE EMISSIONS IN NORTH CAROLINA CROPPING SYSTEMS. A.M. Knight*, W.J. Everman, S.C. Reberg-Horton, S. Hu, D. Jordan, and N. Creamer, North Carolina State University, Raleigh (27)

ABSTRACT

Agriculture accounts for a large portion of land use worldwide. In the U.S. specifically, the World Bank indicated that agriculture accounts for roughly 45% of land use. Agriculture is estimated to contribute greatly to the output of one of the main greenhouse gases, nitrous oxide, which is suspected of contributing to climate change, contributing an estimated 59 percent to emissions. These large percentages are suspected to partially be due to one-third of nitrogen applied to cropping systems being utilized by the system while the additional two-thirds are lost to the environment. With different agricultural practices contributing to these greenhouse gas emissions, finding how various production practices contribute to greenhouse gas emissions will help in the recommendation of best management practices to minimize gas emissions by agriculture in the southeastern U.S. Field studies were conducted in 2013 and 2014 at the Center for Agricultural Farming Systems at the Cherry Research Farm in Goldsboro, NC. Long-term plots of conventional no-till, conventional-tillage, conventional crop-hay, organic tillage, organic minimal tillage, and organic crop-hay systems were used to measure the flux of the greenhouse gases CO_2, CH_4, and N_2O, 24 to 48 hours after ~1.25 cm or more of rainfall, following USDA-ARS GRACEnet Project Protocols. Incubation studies regarding the impact of herbicides on these emissions were conducted in fall of 2014. In these combined experiments it was investigated how weeds and weed control played a role in greenhouse gas emissions. Preliminary results indicated that tillage plays a significant role in gas emissions in cropping systems.

EFFECTS OF A SIMULATED DICAMBA MISAPPLICATION ON NON-TOLERANT SOYBEANS. M.J. Morris*, R.L. Ritter, and D. Hare, University of Maryland, College Park, College Park (28)

ABSTRACT

Dicamba is currently used in many cropping systems for preemergence (PRE) and postemergence (POST) applications. With approval pending for the release of dicamba-tolerant soybeans [*Glycine max* (L.) Merr.], its usage has the potential to increase. Coupled with this increase in usage are the potential hazards of applying the herbicide to non-dicamba resistant soybeans, and the possibility of off-target movement. The objective of this study was to evaluate how a misapplication of dicamba would affect both vegetative and yield components of non dicamba-tolerant soybeans. Field trials were conducted in 2013 and 2014 evaluating eight rates of dicamba (1, 1/2, 1/4, 1/8, 1/16, 1/32, 1/64, and 1/128 pt/A) applied to soybeans at the V3 growth stage. The 1/128 pt/A rate was added to the 2014 trial. A single rate (1 pt/A) of dicamba was also applied at seven soybean growth stages (PRE, VC, V1, V3, V8, R1, and R5). These trials were conducted utilizing Pioneer soybean variety '93Y91,' at Beltsville, MD. Soybean height and stand counts were obtained during the growing season. At maturity, yield, pod count, seed count, and seed weight were obtained. Results from the 2013 study showed that the 1 pt/A rate of dicamba resulted in the most vegetative and reproductive injury while the 1/64 pt/A rate resulted in the least. The 1 pt/A rate was injurious to all growth stages with the exception of PRE and R5 applications. Two greenhouse trials were conducted in 2014 to evaluate soybean response to eleven rates of dicamba (1, 1/2, 1/4, 1/8, 1/16, 1/32, 1/64, 1/96, 1/128, 1/192, and 1/256 pt/A) applied at the V3 growth stage. Plant height, as well as fresh weight and dry weight of the above ground and below ground plant material were obtained. Results indicated that injury became less severe as the rate decreased with the 1/256 pt/A rate being the least injurious.

EVALUATING THE EFFICACY AND FIT OF QUINCLORAC TO CONTROL GRASS WEEDS IN GRAIN SORGHUM IN NORTH CAROLINA. L.J. Vincent*, W.J. Everman, A.M. Growe, T.E. Besancon, A.M. Knight, and Z.R. Taylor, North Carolina State University, Raleigh (29)

ABSTRACT

North Carolina grain sorghum producers are in need of an herbicide component to add to their tank mix to control grass weeds. Quinclorac is known to have a fit in rice and turf production, but is being introduced to sorghum weed management for several reasons. In this particular study we evaluated the impact of quinclorac tank mixed with several products commonly used in sorghum weed management. Data was collected on herbicide efficacy ratings at 7, 14, and 28 days after treatment (DAT), yield, and weed biomass weights and counts.

The protocol included eighteen treatments, half of which were sprayed with a preemergence (PRE) herbicide followed by a post emergence (POST) herbicide application and the other half which received only a POST herbicide application. Both halves had identical POST herbicide applications. The PRE herbicide application was s-metolachlor + atrazine at 1.6 qt. /A. All POST herbicide applications included quinclorac at a rate of 22 fl. oz. /A as well as crop oil concentrate at a rate of 2 pt. /A. In addition to quinclorac and crop oil concentrate at the aforementioned rates, the study treatments included atrazine, pyrasulfotole + bromoxynil, atrazine + pyrasulfotole + bromoxynil, prosulfuron, prosulfuron + atrazine, 2, 4-D, dicamba, and a treatment which omitted a POST application altogether.

End of the year data collection stressed the importance of applying a PRE herbicide to have a successful sorghum crop. When plots which received a PRE followed by POST were compared to plots which only received a POST it was found that weed control of broadleaf signalgrass, large crabgrass and texas panicum was significantly improved. There were no significant differences in control of broadleaf signalgrass, large crabgrass and texas panicum of the PRE followed by POST treatments. However, several significant differences were noted among treatments which only received a POST application. The results from the weed biomass weights and counts reveal that the weight and count of weed species present can be significantly reduced with the application of a PRE herbicide, regardless of whether a POST herbicide was applied. When yield is obtained, it will be used along with weed control rating and weed biomass weights/counts to triangulate winning tank mixes.

GLYPHOSATE RESISTANT PALMER AMARANTH MANAGEMENT IN GLYPHOSATE-TOLERANT SOYBEAN. A.M. Growe*, T.E. Besancon, A.M. Knight, Z.R. Taylor, L.J. Vincent, and W.J. Everman, North Carolina State University, Raleigh (30)

ABSTRACT

The over-exploitation of glyphosate tolerant technology has resulted in increased resistance of troublesome weed biotypes, such as Palmer amaranth (Amaranthus palmeri). As this resistance becomes more common in agricultural systems throughout the southern and mid-western United States, integrated herbicide systems should be practiced to suppress its impact on crop production. Field trials were conducted in Caswell Research station in Kinston, NC and Upper Coastal Plain near Rocky Mount, NC to evaluate the control of glyphosate resistant Palmer amaranth (*Amaranthus palmeri*) with overlapping residual herbicides. Herbicide systems applied PRE were pyroxasulfone, saflufenacil with and without metribuzin, a premix of saflufenacil and imazethapyr with pyroxasulfone, a premix of sulfentrazone and cloransulam alone, and a premix of sulfentrazone and metribuzin alone. These treatments were compared when a POST of Dimethenamid-P was added, with the exception of a sulfentrazone and cloransulam premix. Treatments with Dimethenamid-P as a POST generally were more effective in controlling Palmer amaranth than those without. Between both locations, the PRE only and PRE followed by POST treatments controlled 78 and 87 percent of Palmer amaranth respectively. Control of Palmer amaranth was variable between the two locations. All treatments achieved over 95 percent control at the Caswell station. PRE application of aflufenacil and imazethapyr premix with pyroxasulfone was most effective with 86 percent control of Palmer amaranth at the Upper Coastal Plain location.

HPPD-TOLERANT SOYBEAN SYSTEMS FOR MANAGEMENT OF GLYPHOSATE-RESISTANT PALMER AMARANTH. B.W. Schrage*, M. Rosemond, J. Allen, M.W. Marshall, and W.J. Everman, North Carolina State University, Raleigh (31)

ABSTRACT

The carbon efficiency of Palmer amaranth contributes to its rapid growth, prolific reproduction, and overall competitiveness in North Carolina soybean systems. With the growing presence of glyphosate-resistant biotypes; alternative weed management strategies such as HPPD-tolerant soybeans are being evaluated. An experiment was conducted in Clayton, NC and Blackville, SC in 2014 to assess the efficacy of isoxaflutole and yield in HPPD-tolerant soybeans. Several combinations of isoxaflutole (Balance Pro), flumioxazin (Valor SX), pyroxasulfone (Zidua), and flumioxazin plus pyroxasulfone (Fierce) were applied PRE. Similar POST applications of glyphosate and fomesafen (Flexstar GT) followed at 4 WAP. Plots were rated for percent control of Palmer amaranth at 2, 4, and 7 WAP in South Carolina and 4 and 5 WAP in North Carolina. All plots were harvested upon reproductive maturity. All treatments exhibited greater than 95% control and without phytotoxic symptomology observed on the soybeans; there was no significant difference in yield among treatments in the study conducted in Blackville. In Clayton, treatments failed to display significant differences. This research might suggest the overall effectiveness of proactive weed management efforts to control glyphosate-resistant Palmer amaranth; albeit little difference was noticed among treatments.

IMPACT OF ADAPTIVE MANAGEMENT ON WEED CONTROL IN A LONG-TERM DAIRY CROPPING SYSTEM. K.E. Caswell*, E. Synder, W.S. Curran, H. Karsten, and G.M. Malcolm, Penn State, University Park (32)

ABSTRACT

A long- term, no-till, cropping systems study was established in 2010 seeking to produce enough feed, forage, and fuel to supply the needs of a 65-cow, 240 acre dairy farm in Pennsylvania while minimizing off farm inputs. The study encompasses three unique, six-year crop rotations. Two diverse rotations are compared to a typical no-till corn, soy rotation that uses a typical herbicide regime. One rotation seeks to reduce herbicide use and herbicide resistance through a combination of weed management practices. This rotation is referred to as the Pest Management rotation. In the fourth year of the study, weed management was slightly altered to reflect changes in crop production, study goals, and in response to less effective tactics. Additionally, study management was altered in response to the concerns and interests of a producer advisory panel. The impact of these management changes were evaluated for their ability to control weeds and maintain crops yields, while considering economic returns.

Within the Pest Management Rotation, there are two treatments. The Standard Herbicide (SH) treatment uses a standard herbicide regime for no-till production in a diverse crop rotation. The Reduced Herbicide (RH) regime includes innovative methods that seek to reduce the amount of herbicide use through tactics such as herbicide banding at planting, inter-row cultivation and using a nurse crop in perennial forages. Changes that occurred starting the fourth year include planting soybean in 15-inch rows rather than with a drill in the SH management treatment because of establishment issues. In both the corn and soybean, the RH treatment includes banded herbicide at planting on 30-inch rows and a split treatment of inter-row cultivation or a post-emergence broadcast herbicide. In the perennial forage crops, alfalfa and orchard grass in the RH treatment was initially planted with a nurse crop of triticale and forage peas to help manage weeds. Following continual difficulties with peas, the mixture was reduced to include only triticale as a nurse crop. Herbicide was used to manage weeds in the SH alfalfa monoculture.

Over the first three years, both the SH and the RH management were effective in controlling weeds in both corn and soybean. In corn, the RH management was able to produce comparable yields at equal financial returns to the SH treatment. Challenges in soybean establishment in both the SH and RH managements, along with herbicide selections in RH, made the innovative practices less profitable when compared with SH in the first three years. Management changes starting in 2013 should continue to provide good weed control in both corn and soybean. Changes should improve soybean crop establishment and yield. Altering herbicide selections in both RH and SH beginning in 2013 will allow for a more realistic comparison between the two regimes and three more years of production data will help identify the feasibility of the integrated weed management tactics employed in this project.

IMPACT OF CULTURAL PRACTICES AND WEED MANAGEMENT STRATEGIES ON GRASS CONTROL IN SORGHUM. T.E. Besancon*, A.M. Knight, Z.R. Taylor, L.J. Vincent, W.J. Everman, and R. Weisz, North Carolina State University, Raleigh (33)

ABSTRACT

Weed control remains a major challenge for economically viable sorghum production in North Carolina because sorghum is highly sensitive to weed competition during early growth stages. Moreover, herbicides able to suppress grasses are extremely limited due to sorghum sensitivity. Besides Palmer amaranth (*Amaranthus palmeri*), grasses are the most problematic in sorghum production. Previous studies have shown improvements of weed control in sorghum with narrowed row spacing and increased planting density. Separate field studies were conducted at the Central Crops Research Station (Clayton, NC) in 2013 and at the Upper Coastal Plain Research Station (Rocky Mount, NC) in 2014 to determine which association of row spacing and plant populations would increase crop competitiveness with grasses and eventually allow the reduction of post-emergence herbicide applications. The experiment was conducted in a factorial arrangement of treatments in a randomized complete block design with row spacing (19, 38, and 76 cm), plant population (40,000, 80,000, 120,000, 160,000, or 300,000 plants per acre), and herbicides (non-treated, PRE application of S-metolachlor at 1412 g ai.ha-1 + atrazine at 1824 g ai.ha^{-1}, and PRE application of S-metolachlor at 1076 g ai.ha^{-1} + atrazine at 1390 g ai.ha^{-1} followed by POST application of acetochlor at 840 g ai.ha^{-1} alone or mixed with quinclorac at 289 g ae.ha^{-1}) as main factors. Sorghum was rated for the percentage of large crabgrass (*Digitaria sanguinalis*), crowfootgrass (*Dactyloctenium aegyptium*), broadleaf signalgrass (*Urochloa platyphylla*), and yellow foxtail (*Setaria glauca*) control 4 weeks after PRE, and 1, 3 and 7 weeks after POST. Weed density and biomass were evaluated before harvest as well as yield at the harvest. Data collected stressed the importance of an efficient post-emergence herbicide application in order to successfully control grass species and prevent sorghum yield losses. In 2013 at Clayton, in a situation of low grass infestation, large crabgrass biomass was primarily affected by plant population. Results were similar for crowfootgrass with a decrease of the biomass for narrow row spacing and high plant population. Application of acetochlor as a POST herbicide didn't improve grass control. The highest yields were associated with the combination of narrow rows and high plant densities independently of the herbicide application timing. In 2014 at Rocky Mount, the association of acetochlor and quinclorac applied post-emergence significantly improved grass control in comparison to a single pre-emergence application of S-metolachlor and atrazine. Large crabgrass, broadleaf signalgrass and yellow foxtail were controlled at 100% when a post-emergence application was included in the program versus respectively 92%, 85% and 73% for a single pre-emergence application. Differences in weed biomass according to row spacing and plant population were only recorded for the pre-emergence herbicide application with significant lower grass biomass for the narrow row spacing (7.5 cm) associated with high plant population (120 or 160,000 plants per acre) in comparison to wider row spacing (15 or 30 cm) associated with the lowest plant population (40,000 plants per acre). Highly grass infestation prevented the observation of any significant yield difference according to row spacing or plant population in the non-treated plots.

IMPACT OF TILLAGE ON ITALIAN RYEGRASS CONTROL IN WINTER WHEAT. Z.R. Taylor* and W.J. Everman, North Carolina State University, Raleigh (34)

ABSTRACT

As the occurrence of herbicide-resistant Italian ryegrass (*Lolium multiflorum*) continues to spread, control options are continuing to decline. Cultural practices may become an option to help suppress ryegrass populations and increase control effectiveness when combine with a sound herbicide program. One practice that may have a significant impact on Italian ryegrass population is tillage. To test this theory, wheat was grown in both tilled and in no-till conditions and received the following herbicide treatments; non-treated check, Zidua (pyroxasulfone) at 1.25, and 1.5 oz/a PRE, Fierce (flumioxazin and pyroxasulfone) at 3 oz/a PRE, Zidua at 1.25 oz/a and Sharpen (saflufenacil) at 2 fl oz/a PRE, Prowl H2O (pendimethalin) at 2 pt/a at spike, Axiom (flufenacet and metribuzin) at 8 oz/a at spike, Axiom at 10 oz/a at spike, Zidua at 1.25 oz/a PRE followed by (FB) Osprey (mesosulfuron) at 4.75 oz/a and non-ionic surfactant at 1qt/100gal post, Zidua at 1.25 oz/a PRE FB Powerflex (pyroxsulam) 3.5 oz/a post, Zidua at 1.25 oz/a PRE FB Axial XL (pinoxaden) at 16.4 oz/a post, Osprey at 4.75 oz/a and Zidua at 1 oz/a and non-ionic surfactant at 1 qt/100gal POST, Powerflex at 3.5 oz/a and Zidua at 1 oz/a POST, Axial XL at 16.4 oz/a and Zidua at 1 oz/a POST . In 2013, weed control and yield were both improved in the tilled system compared to the no-till system. Improved control in a tilled system was again seen in two locations in 2014, and yield was improved in one location, while in the other there were no yield differences seen between tillage systems.

IMPROVING CREEPING BENTGRASS RECOVERY FOLLOWING ANNUAL BLUEGRASS CONTROL WITH METHIOZOLIN. K.A. Venner*, S.D. Askew, and S. Koo, Virginia Tech, Blacksburg (35)

ABSTRACT

Methiozolin (PoaCure®), has the ability to completely control annual bluegrass (AB) with safety to creeping bentgrass (CB) when applied at appropriate rates. Sometimes, however, methiozolin removes AB more quickly than desired, resulting in a voided turfgrass canopy. Studies are needed to evaluate methods to increase CB recovery rates when voids are left by dying AB. Studies were conducted to compare increased fertility via soluble fertilizer (SF), SF + trinexapac-ethyl (TE) and Floratine® biostimulants (FB). Two trials were initiated on March 22, 2013 and March 27, 2014. Trial 1 was conducted at the Virginia Tech Golf Course and repeated on two separate, push-up style, practice greens maintained at 0.39 cm. Trial 1 was a randomized complete block design with 3 replications for each study site. Each study site was treated with methiozolin at 3000 g ai ha^{-1} followed by 500 g ai ha^{-1} twice at a 2 week interval in order to facilitate rapid removal of AB. Trial 2 was conducted at the Turfgrass Research Center (TRC) on a USGA specification 'L-93' green that is maintained at 0.32 cm. This trial was aerated to remove 30% of the turfgrass canopy on May 5, 2013 and 2014. Trial 2 was a split-plot design with 4 cultural treatments as main plots and two rates of methiozolin as sub-plots. Sub-plots contained either no methiozolin or methiozolin applied at 500 g ai ha^{-1} 6 times at a 2 week interval. All cultural treatments were the same between trials and were as follows: no cultural treatment, increased fertility using a commercially available SF product, SF plus TE at 0.048 g ai ha^{-1} and increased fertility via FB. All fertility treatments were applied every two weeks, when turfgrass was fully green, beginning on April 14, 2013 and May 2, 2014 in addition to the normal fertility regime administered by a golf course.

Results: Trial 1: At trial initiation, turfgrass was dormant and CB cover ranged from 15 to 21%. One week after initiation of cultural treatments (WAIF) in 2013, FB plots were significantly greener than SF alone and the untreated, but not SF plus TE. In 2014 at 1 WAIF, no differences between treatments was observed. By 6 weeks after initial treatment (WAIT) in 2013, all increased fertility treatments were significantly greener than the untreated, but not different from one another. At 6 WAIT in 2014, SF and FB were significantly greener than the untreated, which was greener than the SF plus TE. At the conclusion of the trial (12 WAIT), cover in treated plots ranged from 95 to 96%, and 90% in the untreated in 2013 and 90 to 93% in 2014.

Trial 2: At trial initiation, turfgrass was dormant, and CB cover ranged from 11 to 21% across the trial area. One WAIF in 2013, percent green tissue on treated plots ranged from 72 to 78%, whereas the untreated was 54%. In 2014, cover at 1 week WAIF ranged from 62 to 70% across all treatments. At 2 weeks after aeration (WAA) in 2013 and 2014, FB recovered more than all other treatments. SF alone increased green cover more than the untreated but not equal to SF plus TE. At the conclusion of the study in both 2013 and 2014, there were no significant differences between cultural treatments. In 2014, however, at 6 WAA until the conclusion of the study, methiozolin treated plots had significantly less green cover than plots that did not receive methiozolin. These data suggest that methiozolin applications could influence turfgrass recovery following an intensive aeration event.

INVESTIGATING THE EFFECTS OF PALE SWALLOWWORT [*VINCETOXICUM ROSSICUM (KLEPOW) BARBAR*] ON ECOSYSTEM FUNCTION. G.L. Thompson* and J. Kao-Kniffin, Cornell University, Ithaca, NY (36)

ABSTRACT

The range of the invasive perennial herb, pale swallowwort (*Vincetoxicum rossicum* (*Kleopow) Barbar.*) is expanding in southern Canada and the northeastern United States. Studies of swallowwort species have indicated phytotoxic effects on co-occurring plants. Furthermore, swallowwort has been shown to affect rhizosphere dynamics through the production of allelochemicals, which have both antifungal and antibiotic properties. We hypothesized that given swallowwort's interference properties, several ecosystem properties mediated by soil microorganisms (such as, organic matter decomposition and nutrient cycling) may be altered when invaded. To test this, we identified five sites around Cayuga Lake, in central New York State, where old fields or woodland edges exhibited areas both invaded and uninvaded by *V. rossicum* for sampling. Sites were sampled during the summer of 2014 consisting of four replicate soil samples per treatment from each site. Within 48-hours of sampling, soils were extracted with KCL to determine potentially available ammonium and nitrate. Fresh soils were also used in soil exoenzyme potential activity assays. Lastly, combined root and soil respiration was measured using a field-based chamber method. No significant differences were found among plant communities for potentially available soil ammonium or nitrate. Additionally, assays of hydrolytic exoenzymes were not affected by invasion, whereas oxidative exoenzymes showed impacts of invasion depending on the site. Soil respiration rates were not altered by pale swallowwort. Together these measurements suggest, upon initial review that structural changes in plant communities due to swallowwort invasion do not lead to large changes in ecosystem processes as found with other invasive species.

IS ECOLOGICAL IMPACT PROPORTIONAL TO COVER? A TEST OF TWO INVADERS. D.R. Tekiela* and J.N. Barney, Virginia Tech, Blacksburg (37)

ABSTRACT

Invasive plant species are synonymous with ecological impact. Unfortunately, these ecological impacts are often assumed and are not empirically measured. Of the studies that do measure ecological impact of invasive plants, very few look at the relationship of invasive plant cover on the magnitude of ecological impact—i.e., is impact proportional with cover? Additionally, the vast majority of impact studies focus on a single invader. However, many ecosystems are experiencing invasion from multiple species simultaneously. Invasive species of similar life history may interact in various ways that may facilitate further invasion – so called invasion meltdown. Here we measured multiple ecological metrics of two invasive grass species, Japanese stiltgrass (*Microstegium vimineum,* JSG*)* and wavyleaf basketgrass (*Oplismenus undulatifolius,* WLBG) across a range of covers to identify the cover-impact relationship and identify their interactions.

We conducted our survey in Soldier's Delight Natural Area within Patapsco State Park where WLBG is thought have initially been introduced in 1996, and where JSG is also abundant. We surveyed 162 1m^2 quadrats systematically randomized within the overlapping range of these two invasive grasses to ensure all combinations of cover of both species were well represented. Within each quadrat the plant community was recorded and soil samples were collected for analysis.

Our results show that biodiversity is strongly negatively correlated with both JSG and WLBG cover. Additionally, increasing cover of JSG reduces plant community richness, though surprisingly, increasing cover of WLBG had no effect on plant community richness. Therefore, the presence of WLBG only reduces evenness and does not impact the total number of plant species present in the forest understory. Furthermore, increasing cover of either invasive grass reduces native plant richness and increases invasive plant richness. Both species modify the structural composition of invaded forests through the reduction of tree seedlings and vining species, but herbaceous cover is not impacted.

Microbial activity was unchanged in the presence of either invasive grass, but JSG cover was correlated with increasing pH. Calcium and magnesium are positively correlated with invasive grass cover but zinc, manganese, copper, iron, boron, and organic matter are not correlated with invasive grass cover.

JSG and WLBG do not modify soil chemistry but do modify the plant community. Most importantly, the presence of either invasive grass reduces tree seedlings, which may alter successional trajectories. These two species may also be shifting the plant community to a greater proportion of invasive species. Both of these outcomes could negatively impact the health of eastern forest ecosystems.

ROUGHSTALK BLUEGRASS CONTROL IN CREEPING BENTGRASS AND KENTUCKY BLUEGRASS FAIRWAYS. S.S. Rana*, S.D. Askew, and S. Koo, Virginia Polytechnic Institute and State University, Blacksburg (39)

ABSTRACT

Roughstalk bluegrass (*Poa trivialis*) is one of the most troublesome weeds of creeping bentgrass (*Agrostis stolonifera*) and Kentucky bluegrass (*Poa pratensis*) fairways. Chemical control options for roughstalk bluegrass are limited in cool-season turf. Methiozolin (PoaCure) is a new isoxazoline herbicide under development by Moghu Research Center that has been reported in Virginia and Korea to control roughstalk bluegrass with safety to creeping bentgrass, perennial ryegrass (*Lolium perenne*), Kentucky bluegrass, and tall fescue (*Schedonorus arundinaceus*). Our objectives were to compare various application timings and tank mixtures of methiozolin for roughstalk bluegrass control in creeping bentgrass and Kentucky bluegrass fairways. Trials were initiated on October 22, 2013 on a creeping bentgrass fairway at the Pete Dye River Course of Virginia Tech (RC), Radford, VA and Highland Golf Course at Primland Resort (PR), Meadows of Dan, VA, and on a Kentucky bluegrass fairway at the Glade Road Research Facility (GR), Blacksburg, VA. Treatments consisted of methiozolin at 1000 or 1500 g ai/ha applied four times at two-week intervals in fall or spring or twice at two-week intervals in fall and spring; methiozolin at 1000 g/ha applied twice at two-week intervals in fall fb by methiozolin plus primisulfuron (26.3 g/ha) or amicarbazone (98 g/ha) applied twice at two-week intervals in spring; primisulfuron or amicarbazone applied twice at two-week intervals in spring; and bispyribac-sodium at 24.7 g/ha (only at creeping bentgrass sites) applied twice at two-week intervals in fall and spring. Treatments also included a non-treated check for comparison. The interaction of location by treatment was insignificant for both roughstalk bluegrass control and turf cover, so data were pooled over locations for mean separation. One year after initial treatment (YAIT), fall and fall fb spring applications of methiozolin at 1500 g/ha controlled roughstalk bluegrass 83-89% and higher than methiozolin at 1000 g/ha at all applications timings, except fall applications of methiozolin fb methiozolin plus primisulfuron in spring. When tank-mixed with primisulfuron, methiozolin at 1000 g/ha controlled roughstalk bluegrass 88% and equivalent to fall or fall fb spring applications of methiozolin at 1500 g/ha. The high rate of methiozolin applied in fall controlled roughstalk bluegrass 89% and higher than the same rate of methiozolin applied in spring. Methiozolin did not injure the desired turf regardless of application rate and timing. Fall only and fall fb spring applications of methiozolin had a turf cover of 80-90% and equivalent to methiozolin tank-mixed with primisulfuron or amicarbazone, but higher than primisulfuron or amicarbazone alone, 1 YAIT. In conclusion, methiozolin at 1500 g/ha controls roughstalk bluegrass better than methiozolin at 1000 g/ha regardless of application timing. Fall applications generally improve roughstalk bluegrass control compared to spring applications only. And, when using the lower methiozolin rate, primisulfuron mixtures help in controlling roughstalk bluegrass equivalent to higher methiozolin rates.

WEED-SOYBEAN COMPETITION IN A LONG TERM ORGANIC CROPPING SYSTEMS EXPERIMENT. M. Ball*, M. Ryan, A. DiTommaso, and L. Drinkwater, Cornell University, Ithaca, NY (41)

ABSTRACT

Weeds are a major challenge in organic grain production. Since total weed control is impractical under organic management, it is important to explore ecological mechanisms that could reduce the negative yield impact of a given weed population. Soil resource partitioning among co-existing species has been documented in natural ecosystems, but has not been explored in detail in agriculture. In this study, we ask whether diversity of soil resource pools can mitigate the negative yield impact of weeds in organic soybean through weed-crop resource partitioning. Our 2014 experiment took place in the Cornell Organic Grain Cropping Systems Experiment (OCS) in Aurora, NY. This long-term experiment compares four organic cropping systems that vary by fertility input, weed management, and tillage regime. We tested soybean biomass and yield response to weed abundance in each system by implementing three weed levels (weed-free, standard management, and supplemented seedbank) in sub-plots within the main system plots. Two additional treatments (+nitrogen and +phosphorous) tested soybean and weed responses to nutrient availability. We compared soil properties between the four cropping systems, including total C and N; mineralizable N; and free and occluded particulate organic matter; to describe soil resource pools potentially relevant to competitive interactions. We also measured soybean nitrogen fixation between cropping systems and weed and nutrient levels. Together, the data give a detailed account of weed-soybean interactions in the varying cropping systems, in which evidence for weed-crop soil resource partitioning can be explored.

EVALUATION OF FALL AND SPRING COMBINATIONS OF PREEMERGENCE HERBICIDES TO PREVENT WEED RESISTANCE IN WILD BLUEBERRY FIELDS.
D.E. Yarborough* and J.L. Cote, University of Maine, Orono (42)

ABSTRACT

In the fall of 2013, a trial was initiated to test the efficacy of several unregistered and/or untested herbicides on wild blueberries (*Vaccinium angustifolium*) in Maine, in conjunction with the industry standards hexazinone and terbacil. Nine sites across the wild blueberry growing region of Maine were sprayed in fall 2013 and/or preemergence in spring 2014 with indaziflam (5 oz/a fall and both in fall+spring; 6.5 oz/a spring), halosulfuron (1 oz/a fall), rimsulfuron (2 oz/a fall) or isoxaben (1.33 lb/a spring). In spring 2014 terbacil (2 lb/a), hexazinone (1 lb/a) or an untreated check was applied at right angles to the test herbicides for a total of 21 treatments. Wild blueberry cover and phytotoxicity, broadleaf weed cover, and grass cover were assessed in June and August 2014 for all treatments as well as the growers' spray regimes outside the trial areas. Soil samples were also collected at each site and analyzed for percent organic matter, soil pH and soil texture. The main effects of the test herbicides alone, with hexazinone, or with terbacil were compared to the untreated check or to the samples from the growers' fields; the test herbicide treatments were also compared to the check and the combinations to the industry standards. One site was dropped because it was over-sprayed by the grower, and six additional sites were excluded after data analysis indicated a lack of weeds across all treatments.

There were no significant differences for main effects of the test herbicides with or without the industry standards, when compared to the untreated check or grower spray regimens. For eight pairwise comparisons, if the variances had been equal there would have been a significant difference, but the unequal variances negated the difference. There were no significant differences in blueberry cover among indaziflam treatments alone, with terbacil or with hexazinone at either assessment. Halosulfuron and halosulfuron+terbacil blueberry cover was significantly greater in August compared to the check or terbacil alone, respectively; otherwise, there were no other significant differences in blueberry cover or phytotoxicity. Phytotoxicity was minimal in general, and none was noted at the August assessment. The indaziflam, halosulfuron and isoxaben treatments combined with terbacil suppressed broadleaf weeds significantly more than terbacil alone. Indaziflam+terbacil also significantly suppressed broadleaf weeds compared to the check in June, but not in August. Grasses were essentially eliminated by terbacil and the terbacil combinations; otherwise, although grass pressure was very low there were no differences among the check and the herbicides alone, or hexazinone and its combinations. In summary, indaziflam, in combination with terbacil, is effective in controlling both broadleaf weeds and grasses whether applied in the fall or spring. Isoxaben also appears effective on both broadleaf weeds and grasses in combination with terbacil; halosulfuron alone or with terbacil was effective on grasses, but in this trial did not offer long-term suppression of broadleaf weeds. Rimsulfuron exhibited fair control of weeds, but was not significantly more effective than the standards alone. The combinations of fall and spring treatments provided for improved weed control versus what growers are currently using.

MANAGEMENT OF PERENNIAL SOWTHISTLE IN DRY BULB ONION IN NEW YORK STATE. C.A. Hoepting* and E.M. Buck, Cornell University Cooperative Extension, Albion, NY (43)

ABSTRACT

Perennial sow thistle (*Sonchus arvensis L.*) has emerged as a devastating weed problem in certain muck land areas where onions (*Allium cepa* L.) are grown in New York. When uncontrolled, it can significantly reduce onion yield by 84 percent and drastically reduce bulb size. The herbicides commonly used in muck onion production do not provide any control of this weed and hand weeding and cultivation can exacerbate the problem. In this project, we identified how to effectively manage perennial sow thistle with Stinger (ai. clopyralid) in onions.

Trials were conducted in commercial muck fields of direct seeded yellow dry bulb onions in 2013 ('Centerstone') and 2014 ('Talon'). In 2013, maximum rate of 0.375 lb ai/A of clopyralid treatments (in fl oz:"Max 16") included 1 by 0.75 lb ai/A, 2 by 0.187 lb ai/A, 4 by 0.094 lb ai/A, 0.187 followed by 2 by 0.094 lb ai/A (in fl oz: 8-4-4), and 0.094 followed by 0.187 and 0.094 lb ai/A (in fl oz: 4-8-4). In a crop tolerance trial in 2013, single applications of 0.094, 0.187 and 0.375 lb ai/A were applied to onions at the 4-, 5-, and 6-leaf and 1-inch bulb stages; 0.094 and 0.187 lb ai/A were also applied to 2-leaf onions. Additionally, 4 by 0.094 lb ai/A, 2 by 0.187 ai/A, 8-4-4 and 4-8-4 programs were applied starting at the 2- and 4-leaf stages. In 2014, Max 16 treatments included 8-4-4, 6-6-4 (6 fl oz = 0.141 lb ai/A) and 12-4 (12 fl oz = 0.281 lb ai/A), which began at 2-leaf onion stage when Perennial sow thistle was at the mid-rosette stage. Clopyralid 0.187 (4-leaf) + 0.094 lb ai/A (5-leaf) was applied: by itself and followed by hand weeding, following hand weeding (1-leaf onions), following flumioxazin 0.064 lb ai/A (2.5-leaf onions), and following ethofumesate 0.5 lb ai/A (2.5-leaf onions). Finally, clopyralid 0.094 and 0.187 lb ai/A was applied to 2.5- and 5-leaf onions, respectively.

The most susceptible stage of perennial sow thistle to chlopyralid was the mid- to late-rosette stage which has 5 to 9 expanded leaves (4 to 7 inches in diameter), while newly emerged and bolt stages were the most difficult to control. When timed to this weed stage, a single application of 0.187 lb ai/A achieved best control compared to 0.141, 0.094 and 0.281 lb ai/A and control was consistently improved with multiple applications. Of the "Max 16" treatments, best control was achieved with 8-4-4, which resulted in 99 percent control of weed biomass. Onions could not tolerate chlopyralid after they started to bulb, especially when 0.187 lb ai/A or more was applied. Also, bulb injury increased when 0.187 lb ai/A or more was applied to 2-leaf onions. Single applications of 0.091, 0.187 and 0.375 lb ai/A to onion at the 4-, 5- and 6-leaf stages resulted in less than 5% unmarketable bulbs. In 2014 trials, none of the clopyralid treatments totaling 0.281 lb ai/A made in two applications between the 4- and 6-leaf stages resulted in more than 1.1% unmarketable bulb injury. Ultimately, clopyralid 0.187 lb ai/A timed to the susceptible weed stage followed by 0.094 lb ai/A 1 to 2 weeks later when injured weeds begin to recover will be recommended. Since the susceptible weed stage generally occurs when the onions have only 2 leaves, effective strategies to align susceptible weed stage with most tolerant onion stage (4-leaf) included hand weeding, and application of flumioxazin or ethofumesate herbicides. Alternatively, clopyralid 0.091 lb ai/A could safely be applied to 2-leaf onions timed to susceptible weed stage and then followed with 0.187 lb ai/A at the 4-leaf stage.

BICYCLOPYRONE FOR WEED CONTROL IN VEGETABLE CROPS. B. Zandstra* and C.J. Phillippo, Michigan State University, East Lansing (44).

ABSTRACT

Experiments were conducted to determine crop safety and potential uses for bicyclopyrone in vegetable crop production. Bicyclopyrone is an HPPD inhibitor, which inhibits carotenoid synthesis, resulting in bleaching in susceptible plants. Bicyclopyrone was applied at rates of 0.037 and 0.05 kg/ha preemergence and postemergence on various vegetable crops. Crops were evaluated for visual injury and yield. Weeds were rated visually for control. Bicyclopyrone was applied preemergence at 0.05 kg/ha to third-year asparagus (*Asparagus officinalis* L.) It caused slight stunting early in the season but yields were not reduced. Bicyclopyrone was safe on carrot (*Daucus carota* L.) when applied preemergence at 0.037 or 0.05 kg/ha. When applied postemergence to carrot, yield was suppressed slightly. On onion (*Allium cepa* L.), bicyclopyrone at 0.05 kg/ha was safe preemergence with no yield reduction. It caused serious onion injury when applied postemergence. Seeded pumpkin (*Cucurbita pepo* L.) was moderately tolerant of bicyclopyrone, but butternut (*Cucurbita moschata* Duch. Ex Poir.) and buttercup (*Cucurbita maxima* Duch.) squashes were stunted. Bicyclopyrone stunted established chives (*Allium schoenoprasum* L.) moderately. Bicyclopyrone applied pretransplanting was safe on cabbage (*Brassica oleracea* L. var. *capitata* L.) and Chinese cabbage (*Brassica rapa* L. var. *pekinensis* (Lour.) Olsson. It caused serious crop injury when applied posttransplanting to broccoli (*Brassica oleracea* L. *italica* Plenck.), cabbage, and Chinese cabbage. Bicyclopyrone applied postemergence 30 days after planting was safe on broccoli and cabbage but killed Chinese cabbage. It caused unacceptable injury when applied pretransplanting to pepper (*Capsicum annuum* L.) and tomato (*Lycopersicon lycopersicum),* and preemergence to seeded basil (*Ocimum basilicum* L.), dill (*Anethum graveolens* L.), fennel (*Foeniculum vulgare* var. *azoricum* (Miller) Thell.), lettuce (*Lactuca sativa* L. var. *capitata* L.), and parsley.

Bicyclopyrone controlled barnyardgrass (*Echinochloa crus-galli* (L.) Beauv.), green foxtail (*Setaria viridis* (L.) Beauv.), and large crabgrass (*Digitaria sanguinalis* (L.) Scop.) preemergence. It did not control field sandbur (*Cenchrus incertus* M.A. Curtis). It controlled common lambsquarters ((*Chenopodium album* L.), common ragweed (*Ambrosia artemisiifolia* L.), eastern black nightshade (*Solanum ptycanthum* Dun.), and redroot pigweed (*Amaranthus retroflexus* L). It controlled ladysthumb (*Polygonum persicaria* L.) preemergence on mineral soil but not on organic soil, and controlled it postemergence on both soils. It did not control common purslane (*Portulaca oleracea* L.). Bicyclopyrone may be an effective herbicide for annual grass and broadleaf weed control in some vegetable crops.

USE OF ADJUVANTS PRIOR TO POST APPLICATIONS OF OXYFLUORFEN INCREASE INJURY IN BROCCOLI AND CABBAGE. R. Bellinder* and C.A. Hoepting, Cornell University, Ithaca, NY (45)

ABSTRACT

A grower in western New York reported in 2013 that he had gotten severe crop injury in his direct-seeded broccoli following postemergence applications of oxyfluorfen. New York is one of only two states to have obtained a 24c Special Local Need (SLN) registration for this use. Prior to this only slight necrotic spotting had occurred in field trials. The grower followed the rules on the label with regard to rates, weather conditions, and did not use any adjuvants. Discussions with the grower revealed that he had applied the insecticide, Coragen, with a highly effective adjuvant, NuFilm P one and eight days prior to applying oxyfluorfen in two separate fields. The crop that was treated one day following the insecticide did not recover whereas the one that was treated eight days later did outgrow the injury.

During 2014, trials were conducted in transplanted broccoli and cabbage at the HC Thompson Vegetable Research Farm in Freeville, NY.to evaluate the impact of applying adjuvants with and without Coragen on crop safety. The adjuvants used were Activator 90 NIS, Agri-Dex COC, and Nu-Film P. All three were applied alone or with 0.065 lb ai/A Coragen 19 days after transplanting. Oxyfluorfen at 0.188 lb ai/A was applied either 1 or 6 days following the adjuvant/Coragen treatments. Overall broccoli was more sensitive than cabbage and when oxyfluorfen was applied 6 days after the adjuvants it caused virtually no injury in either crop. Injury following the three adjuvants when applied alone was slight. When the Coragen was added necrosis and epinasty increased significantly with all three adjuvants. These treatments continued to show some injury when evaluated 20 days after the early treatment in both crops. Numbers of heads and total weights were highly variable and were reduced only in cabbage when the Coragen was applied with NIS and COC.

THE IR-4 PROJECT: UPDATE ON WEED CONTROL PROJECTS (FOOD USES). M. Arsenovic*, D. Kunkel, and J. Baron, IR-4 Project, Princeton, NJ (46)

ABSTRACT

The IR-4 Specialty Crop Program is a publicly funded program that develops and submits regulatory data for the registration of pest control products on specialty crops. IR-4 has a long history of providing herbicide registrations for specialty or minor crop growers. In 2014, IR-4 data was used to support a number of new herbicide registrations. These included new registrations for linuron, clomazone, and fenoxaprop-ethyl with others to follow in 2014, such as new uses for sethoxydim. A large number of herbicide submissions were made to EPA in 2014. These included S-metolachlor for lettuce and other crops, prohexadione calcium for strawberry, ethofumesate to reduce plant back restrictions, halosulfuron-methyl on a number of fruit crops, pronamide re-activation for leaf lettuce, with others submissions to follow by the end of 2014, such as carfentrazon-ethyl, clethodim, fluaziflop-p-butyl and penoxsulam. The 2015 IR-4 research plan will likely include close to 40% herbicide projects. The new IR-4 strategic plan (2015-2020) was approved in 2014 The IR-4 Project will remain a responsive and efficient organization that supports US stakeholders by facilitating U.S. regulatory approvals for crop protection tools. IR-4 will continue with a research focus on low risk pesticide registrations that support integrated pest management systems. There will likely be more emphasis on efficacy and crop safety data to determine products that can manage hard to control pests and mitigate invasive species. There will be continued support for products that can be used for organic growers as well as cutting edge biotechnology products. Other areas expected to have greater importance will be in the area of international harmonization, to assist in exports of U.S. grown specialty crops and removing pesticides as a technical trade barrier.

INTRODUCING THE USE OF ANALYTIC HIERARCHY PROCESSING FOR DECISION MAKING IN WEED MANAGEMENT. H. Sandler*, K.M. Ghantous, and K. Hansen, UMass Cranberry Station, East Wareham, MA (47)

ABSTRACT

Analytical Hierarchy processing (AHP) can be described as a mathematical model of multi-criteria decision-making (MCDM) that uses pairwise comparisons, in conjunction with a ratio scale, to indicate the strength of preference or importance of various criteria. AHP is a powerful and flexible method for decision-making. AHP incorporates the values of the decision maker and stresses the importance of these intuitive judgments; it also calculates an index regarding the consistency of the judgments. AHP enables users to set priorities and make decisions when both quantitative and qualitative elements need to be considered. The method has been used for a wide range of issues in business, teaching, and politics for topics such as selecting the best health care service, prioritizing academic initiatives, and determining the best renewable energy resource. Its adaptation into the realms of biology and agriculture has been slow but AHP has been used in various aspects of environmental management to validate information for projects such as risk assessment for loggerhead turtles and prioritizing risks facing cocoa farmers.

A project to design and implement AHP to facilitate decision-making for weed control in cranberry production in Massachusetts is underway. Dodder (*Cuscuta gronovii* Willd. ex Schult.) is one of the most problematic weeds in cranberry production and can cause significant yield reduction. Managing dodder is not straightforward as no single tactic provides adequate control across years and locations. An integrated approach must be used and often, a different approach may be needed for the same grower in one year as compared to the next. It is often a difficult, confusing, and sometimes arbitrary process to decide what is important and what tactic should be used in any given year.

AHP is especially attractive for this particular weed problem because it organizes both tangible (e.g., chemical efficacy) and intangible factors (e.g., a grower's experience or preference) in a systematic way. Using SimpleMind mapping software (ModelMaker Tools BV, Netherlands), the important criteria that growers must consider when making management decisions for dodder control have been represented in an easy-to-understand flow-chart type diagram. The maps were used to generate ratio-scaled surveys specifically designed so the results can be entered into a spreadsheet containing the formulas to prioritize the selected criteria. Over the next few months, we will be pilot-testing various AHP scenarios to evaluate if AHP is a reasonable approach for simplifying dodder management decisions and to gauge if growers would be willing to use this technique in their weed management program.

NON-NATIVE FLORA OF SIX NEW YORK ISLANDS. R. Stalter*, St. Johns University, Queens, NY (48).

ABSTRACT

The objective of the present study was to compare non-native and native species richness at six New York coastal islands. These islands listed in descending order by size are Ellis (11.l ha) Great Gull (6.9) Liberty (4.9) Hoffman (4.0) Swinburne (1.0) and Pea (0.8). The vascular plant species at all six islands has been collected and identified by the author. The greatest number of taxa occurred at Ellis Island, the largest island with 247 taxa while man-made Swinburne Island had the fewest, 38 taxa. Two uninhabited islands, Hoffman and Swinburne located 1.5 km east of Staten Island were populated with 87 and 38 taxa respectively. Hoffman and Swinburne islands were constructed in the late 1800's to temporarily house immigrants denied entry to the United States.

Liberty (97 taxa) and Ellis (247 taxa) islands located in New York City harbor are visited daily by thousands of tourists. Great Gull Island (203 taxa) located 3 km east of Orient: Point the tip of Long Island is the site of the American Museum of Natural History's tern monitoring study is visited by researchers in the spring nesting season each year. Ellis and Liberty islands are maintained by ground keepers and inundated daily with hundreds of visitors while Great Gull Island Islands is bulldozed and brushed hogged to provide favorable nesting habitat for nesting terns each year. Hoffman and Swinburne located one km east of Staten Island are uninhabited and only occasionally visited by National Park Service personnel. Pea Island (68 taxa) located 1.5 km south of New Rochelle in western Long Island Sound site of the abandoned Huguenot Yacht Club. Pea Island has been devoid of human presence since the site was savaged by the December nor'easter of 1992.

There is a significant difference in the percent of non-native taxa found at Pea Island, 43%, with the percent of non-native taxa present at the other five islands where non-native taxa ranged from 57% at Great Gull to 65% at Liberty Island. Island size, geology, origin, disturbance and human manipulation at Ellis, Great Gull and Liberty contributes to the large percentage of non-native taxa at these islands

Table 1. Frequencies of native versus non-native plants at New York State, Pea Island, Liberty Island, Ellis Island, Hoffman Island, Swinburne Island and Great Gull Island.

	New York State	Pea Island	Liberty Island	Ellis Island	Hoffman Island	Swinburne Island	Great Gull Island
Native species	2078	39	34	98	31	14	73
Non-native species	1117	29	63	149	56	24	115
% Non-native species	35.0	43.0	65.0	60.3	62.0	63.0	57.0
Total species	3195	68	97	247	87	38	203

WEED AND COVER CROP BEHAVIOR IN AN INTER-SEEDED COVER CROP PLUS REDUCED-RATE HERBICIDE MANAGEMENT SYSTEM IN WIDE-ROW VEGETABLE CROPS. V. Bhaskar*, R. Bellinder, A. DiTommaso, and M. Walter, Cornell University, Ithaca, NY (49).

ABSTRACT

Over the past few decades, studies on cover crops have been extensive. The impacts of inter-seeded cover crops on weed growth, cover crop and cash crop performance has also received considerable attention. Information on living mulch cover crops managed with herbicide application, especially in less mechanized and more rigorous systems like vegetable crops, has been more limited. Direct competition with cash crops and unreliable weed control are the main reasons interest in inter-seeded cover crops in high value crops has been low. Thus, there is a need to assess additional cover crop species for inter-seeding that not only have the capacity to smother weeds but that can also produce high biomass and have favorable growth habits for their management.

In the summer of 2014, a field experiment was performed at the Homer C. Thompson Vegetable Research Farm in Freeville, NY, to study the performance and potential for inter-seeding of two tropical cover crop species, sesbania (*Sesbania sesban*) and sunnhemp (*Crotalaria juncea*), in wide-row transplanted tomatoes. The trial was set up in a split plot design with the two cover crop species as the main plots and different management methods as sub plots. Both sesbania and sunnhemp were planted in rows 20cm apart. Tomato plants (Mountain Fresh F1) were transplanted at 1.22m (4') between-row spacing and 0.46m (1.5') plant-to-plant spacing by removing cover crop rows so that tomato rows were 40cm from the closest cover crop row on either side. There were 3 rows of cover crop between 2 rows of tomatoes. The treatments within the cover crop blocks were a handweeded check, mowing, two rates of metribuzin (0.136 and 0.21 kg ai/ha in sesbania and 0.08 and 0.136 kg ai/ha in sunnhemp) and rimsulfuron (0.0068 and 0.011 kg ai/ha in sesbania and 0.0085 and 0.017 kg ai/ha) and a single rate of halosulfuron (0.045 kg ai/ha in sesbania and 0.053 kg ai/ha in sunnhemp). Cover crop performance (biomass and ground cover) and weed cover were recorded thrice during the season and tomato yield from each plot was recorded.

Due to the unusually cold summer, initial cover crop growth was very slow. This was more evident in sesbania, which performed very poorly with less than 0.5 tons/ha of fresh biomass; sunnhemp produced an average of 4 tons/ha. Preliminary analysis of this season's data showed that cover crop had no effect on tomato yield but that management treatments did have a significant impact. Highest tomato yields were recorded in plots subjected to high rates of metribuzin (78 tons/ha, averaged across both cover crop species) and lowest in the mowing treatments (42 tons/ha). These differences were most likely due to weed presence since the lowest (<10% on average across the two species) and highest weed covers (42%) were measured in the high-rate metribuzin and mowing treatment plots, respectively. High-rate rimsulfuron and low-rate metribuzin also had yields comparable to high-rate metribuzin plots and were higher than the control plots (68 tons/ha). The high rate of rimsulfuron showed promise in temporary arrest of sunnhemp growth.

TESTING THE CONSEQUENCES OF FUNCTIONAL TRAITS ON LANDSCAPE SPREAD USING A SIMULATION MODEL. J.N. Barney*, D.R. Tekiela, and E. Dollete, Virginia Tech, Blacksburg (50).

ABSTRACT

The spatial distribution of plants at the landscape level is a function of the dispersal dynamics of the plant, population growth rate, plant functional traits, and landscape connectivity. Much work has been done to understand the dynamics of plant populations, including determining their population growth rates. Less is known about how plant populations fair under a variety of habitat conditions. For example, we would not expect the same population growth rates of a species in a pasture, deciduous forest, and riparian edge. Similarly, a variety of models have been developed to simulate landscape level population dynamics. However, the majority of these models use simulated landscapes, or assume equivocal invasibility across diverse landscapes. Therefore, to help bridge the disparities among existing models, we have developed a stochastic simulation model that integrates species functional traits, dispersal kernels, and population dynamics onto real landscapes that can vary in their susceptibility and resistance to invasion.

The InvDER model—Invader Disperse/Establish Risk model—is a flexible platform to model any species across any landscape spanning a specified timeframe. For this study we varied species functional traits and population size and location to identify their role in landscape level spatial processes. We compared annual vs perennial plants, keeping all other parameters constant. Perennial wind dispersed plants spread faster and had larger final population sizes than equivalent annuals. Not surprisingly, wind dispersed plants spread faster and had greater final population sizes than equivalent gravity dispersed plants. The size and distribution of the initial invasion had non-linear effects on the size of the final invasion.

The InvDER model allows the testing of various aspects of the invasion process, from functional traits to landscape diversity and composition that can facilitate conceptual development as well as foster effective and targeted management.

DOES METHODOLOGY MATTER? A STUDY OF HOW METHODS INFLUENCE INVASIVE PLANT IMPACT RESEARCH. D.R. Tekiela* and J.N. Barney, Virginia Tech, Blacksburg(51).

ABSTRACT

Invasive species are often assumed to have negative ecological impacts on the ecosystems they invade, yet only recently has the field of invasion ecology begun to empirically study these impacts. Unfortunately, these impact studies are fragmented and several methodologies exist for determining impact. The impact of invasive species cannot be determined without a comparison to a benchmark from which to identify a change, and not all studies use similar points of comparison. Methodology is divided into three primary categories of benchmarks: 1) uninvaded reference sites; 2) invader removals; 3) and invader addition. In all instances the invaded sites is compared to the benchmark. Each benchmark has advantages and disadvantages. Uninvaded sites are assumed to be susceptible to invasion, but may differ and not be invasible. Removing invaders assumes that invaders are the drivers of ecological impact, and that this impact will return immediately following removal, and return to the preinvaded state. Adding the target invader assumes impacts accumulate immediately, which is largely untested. Because no standard exists, it is assumed that these benchmarks yield similar results, but this has never been empirically tested. Therefore, our objective is to determine if impacts are dependent on benchmark methodology.

To determine the effect of methodology on impact results, we have created a database of invasive species impact studies. This database consists of 113 papers. There is an even distribution of studies across the potential range of percent cover and of these studies 58% use an uninvaded benchmark, 40% use invader removal, and 2% added the invader. The greatest proportion of studies were performed in forest systems (17.5%) followed by riparian systems (12.6%). The most commonly studied species is *Bromus tectorum* (downy brome; 16%), followed by *Ligustrum sinense* (Chinese privet; 9.7%), and *Phragmites australis* (common reed; 7.8%). The most common type of metrics measured were plant community metrics (47.8%), and within plant metrics plant community richness is the most common metrics measured (27.5%) followed by plant community diversity (18.3%).

Initial testing suggests that the methodology used to determine impact does impact the conclusions of impact studies. Removal studies conclude larger negative impacts then observational studies. Therefore, our results suggest that a unifying methodology should be developed for future invasive plant impact studies. Future work aims to control for ecosystem type, plant life form, cover level, and species in testing these differences

EVALUATING TOLERANCE OF INTER-SEEDED COVER CROPS TO PRE-EMERGENT HERBICIDES FOR CORN PRODUCTION IN THE MID-ATLANTIC. J.M. Wallace*, W.S. Curran, M. Ryan, S. Mirsky, and D. Lingenfelter, Pennsylvania State University, University Park (52).

ABSTRACT

Integrating cover crops into annual crop rotations can improve soil quality, nutrient cycling, pest regulation and crop productivity. Cover crop establishment in no-till corn production is often constrained by a short cover crop growing season following corn harvest within the Mid-Atlantic region. Relay cropping, or interseeding, can facilitate the establishment of cover crops by sowing cover crops into standing corn. Recent field studies suggest that annual ryegrass and several leguminous species can be established into corn at the V5 growth stage, approximately 35 days after corn planting. At this timing, cover crop establishment may be exposed to soil residual herbicides applied at corn planting. In this study, we evaluated the effects of various pre-emergent herbicides on interseeded cover crops in no-till corn production.

We conducted a field trial in 2013 and 2014 at the Rock Springs Agronomy Farm in Centre County, PA on a Hagerstown silt loam. Field trials were designed as a split-plot with four replications. Whole plots (3 × 24 m) consisted of eleven different pre-emergent herbicides plus an untreated check applied at a single rate perpendicular to corn rows. Herbicide treatments included: s-metolachlor, dimethenamid-P, acetochlor, pendimethalin, pyroxasulfone, saflufenacil, rimsulfuron, atrazine, metribuzin, isoxaflutole and mesotrione. Splits plots consisted of four cover crop species interseeded in eight corn rows across main plots. Cover crop species included annual ryegrass seeded at 22.4 kg ha^{-1} and medium red clover, crimson clover and hairy vetch at 11.2 kg ha^{-1}. Pre-emergent herbicides were applied following planting of glyphosate-resistant corn in late-May each year. Glyphosate was applied across all plots approximately 28 days after planting and cover crops were interseeded 35 days after planting. Treatments were visually evaluated (% stand reduction) 30 days after interseeding and aboveground biomass was harvested in late fall. Rainfall was near normal in 2013 and results showed that mesotrione significantly reduced red clover and hairy vetch biomass in comparison to the untreated check. Pyroxasulfone and s-metolachlor applications significantly reduced annual ryegrass biomass in comparison to the untreated check. Cover crop biomass was similar to the untreated check in all other herbicide treatments.

We initiated a second field trial in 2014 at Rock Springs PA, Landisville PA, Ithaca NY and Beltsville MD to examine herbicides we believe have potential for success with interseeding. Whole plots (3 × 8 m) consisted of broadleaf (atrazine, saflufenacil, rimsulfuron) and grass (pendimethalin, dimethenamid-P, acetochlor) herbicide treatements applied at a ½ and 1X label rate and broadleaf + grass (acetochlor + atrazine, pendimethalin + atrazine, acetochlor + rimsulfuron, dimethenamid-P + saflufenacil) herbicide treatments. Split plots consisted of annual ryegrass (22.kg ha^{-1}) and medium red clover (22.4 kg ha^{-1}) interseeded in two corn rows per main plot. The experiment was otherwise managed using methods from the previous study. Results of the PA field trials show that herbicide treatments did not reduce annual ryegrass and red clover biomass in comparison to the untreated check on our silt loam soil. In addition, no differences were found between ½ and 1X rate treatment comparisons. Rainfall in 2014 was again plentiful. In summary, the preliminary results of our field trials suggest that pre-emergent residual herbicide programs can be designed for corn production systems that result in negligible to limited injury to interseeded grass or leguminous cover crops. However, soil texture, pH and seasonal rainfall will influence herbicide persistence, so our preliminary results should be used with caution.

COVER CROP ADOPTION AND UTILIZATION IN PENNSYLVANIA; A COMBINATION OF STAUNCH ENTHUSIASM AND ANXIOUS UNCERTAINTY. W.S. Curran* and D. Lingenfelter, The Pennsylvania State University, University Park (53).

ABSTRACT

Cover crop enthusiasm in the mid-Atlantic region has grown rapidly. Both researchers and farmers are testing new ways to establish and manage cover crops in no-till systems. The most frequent questions asked to regional weed science extension specialists include concerns about appropriate herbicides that will allow successful cover crop establishment on the front end and how to kill or control cover crops on the back end before no-till establishment of cash crop. We have had several studies focused in both of these areas.

Our herbicide carryover research to date suggests that most of the herbicides used in corn and soybean are relatively safe to common fall seeded cover crops. The exception to this is some concern to small seeded legumes and *Brassica* species from selected herbicide products. The type of research necessary to determine crop safety can be time consuming mostly because the research needs to be conducted at multiple locations and perhaps over multiple years to have the confidence necessary to make sound recommendations. As an alternative, we have focused our efforts on using herbicide half-life information published in the scientific literature along with cover crop sensitivity or susceptibility in explaining the potential impact of herbicides on cover crop establishment. A major limiting factor with this approach is reliance on half-life information which can vary quite widely depending on soil and environmental factors. However, we believe this approach is conservative and will likely over predict potential problems, but this is preferable to the alternative.

Herbicide applicators have also expressed concern about adequate control of certain cover crops and especially cereal rye and annual ryegrass in the spring in their burndown program. Our cover crop control experiments have routinely showed that herbicides can effectively control grass and legume cover crops with few problems assuming the correct product and rate are selected. We have concluded that greater attention to detail at application time that includes using the appropriate carrier volume, herbicide, herbicide rate, and avoiding certain tank-mixtures can eliminate or reduce the potential for poor herbicide performance.

SWEET CORN AND POPCORN TOLERANCE TO POSTEMERGENCE APPLICATIONS OF NICOSULFURON PLUS MESOTRIONE. D.D. Ganske*, G. Hannig, V. Kleczewski, H.A. Flanigan, and J. Bugg, Dupont, Winchester, VA (54).

ABSTRACT

The use of herbicide resistant crops has become very widespread during the last ten years, in part, due to their role in easier and cheaper weed management programs. However, in the specialty corn market, the use of herbicide resistant traits is not an option and therefore conventional herbicides are needed for weed management. DuPont Crop Protection, under the trade name of Revulin ™ Q, has developed a formulated premix of nicosulfuron + dry mesotrione + isoxadifen-ethyl crop safener for postemergence application to sweetcorn, popcorn, and seed corn that provides selective postemergence control of tough grass and broadleaf weeds. This premix combination also offers two modes of action, a low use rate, a convenient dry formulation, and excellent crop safety. The addition of the crop safener provides the flexibility for Revulin™ Q to be applied under more diverse weather conditions and across a wide application window to more seed corn inbreds, sweet corn and popcorn hybrids by reducing the risk of crop injury. The objective of this study was to evaluate Revulin™ Q in one pass and two pass weed management programs for weed control and crop response on sweetcorn and popcorn hybrids. The experiment was conducted as a randomized complete block design with 3 to 4 replications using 11 popcorn hybrids and 17 sweetcorn hybrids at various locations in the United States. Weed control and crop response were evaluated on a scale of 0 =no injury or control and 100% = complete kill at 7 and 14 Days After Treatment (DAT). Difficult to control grass and broadleaf weed species such as large crabgrass, fall panicum, ragweeds, and pigweeds were present in the trial locations. At 7 DAT, Revulin™ Q applied to popcorn showed only 2% injury on average when applied alone and 8% injury when tank mixed with other herbicides. Similar results were observed with sweetcorn at 7 DAT showing only 5% injury when applied alone and up to 8% injury when tank mixed with other herbicides. Injury decreased on both crops to 6% or less for all herbicide combinations at 14 DAT. Overall, Revulin™ Q applied postemergence to popcorn and sweetcorn was found to be safe with a low risk of injury across all hybrids tested. Grass and broadleaf weed control was greater than 90% for many of the weeds evaluated, including: large crabgrass, barnyardgrass, green foxtail, common lambsquarters, and common ragweed. The addition of Revulin™ Q to the specialty corn market will provide an effective conventional herbicide option with two modes of action and excellent crop safety.

CONTROL OF ALS-RESISTANT COMMON CHICKWEED. E. Hagood and M.L. Flessner*, Virginia Tech, Blacksburg (55).

ABSTRACT

Common chickweed (*Stellaria media* (L.) Vill.) control failures were observed following application of thifensulfuron + tribenuron, two ALS-inhibiting herbicides, in the 2007-08 wheat production season in New Kent (NK) and King William (KW) counties in eastern Virginia. Herbicide resistance was suspected and research was conducted across three objectives: 1) confirm ALS-resistance, 2) evaluate possible cross resistance, and 3) assess alternative control options in the field. For objectives one and two, seed was collected from four control failure sites (NK1, NK2, KW1, and KW2). An ALS-susceptible wild type (WT) population was collected in Montgomery county Virginia. Seed were germinated and established in 5.5 cm^2 pots with commercial potting soil in the greenhouse. Herbicide treatments were applied to 3 to 4 cm tall common chickweed in 230 L ha-1 spray at 270 kPa. Four treatment replicates were conducted and the experiment was repeated. Thifensulfuron was applied at 0, 0.25, 0.5, 1, 2, 4, 8, 16, and 32X the labeled rate of 0.026 kg ai ha-1. F2 generation seed was collected for the NK1 population, and the experiment was repeated to assess resistance heritability. For objective two, common chickweed seedlings were treated with one and four times the labeled rate of metsulfuron-methyl, prosulfuron, imazethapyr, imazapyr, and bispyribac, which were 0.013, 0.020, 0.070, 0.140, and 0.074 kg ai ha-1. For both objectives one and two, vigor reduction (% reduction) was assessed 28 days after treatment (DAT) and fresh weights were evaluated 30 DAT. For objective three, field experiments were conducted from 2008-14. Herbicide treatments were applied as previously described to 3 by 7.6 m plots and were replicated a minimum of three times. Preemergence (PRE) treatments included flumioxazin at 0.036, 0.07, and 0.14 kg ai ha-1. Delayed PRE (DP) treatments were applied to one to two leaf wheat with no germinated common chickweed and included pendimethalin at 1.06 kg ai ha-1, metribuzin at 0.05 or 0.11 kg ai ha-1, and flufenacet + metribuzin at 0.23 + 0.06 kg ai ha-1. Postemergence (POST) treatments were applied to common chickweed with six to eight leaves and four cm in height and included thifensulfuron + tribenuron at 0.13 and 0.06 kg ai ha-1, 2,4-D + dicamba at 0.56 + 0.14 kg ai ha-1, pyrasulfotole + bromoxynil at 0.03 + 0.24 kg ai ha-1, fluroxypyr at 0.11 and 0.14 kg ai ha-1, and metribuzin at 0.11 and 0.21 kg ai ha-1. Other trials assessed thifensulfuron + tribenuron at 0.018 and 0.009 kg ai ha-1 tank-mixed with metribuzin at 0.11, 0.16, or 0.22 kg ai ha-1 or fluroxypyr at 0.10 or 0.12 kg ai ha-1 applied POST fall, POST early-spring, or POST late-spring. Visible percent common chickweed control data were collected 28 DAT. For all objectives, herbicide treatments were applied with nonionic surfactant at 0.25 % v v-1, and data were subjected to ANOVA and means separated (LSD, Î± = 0.05) accordingly. Vigor reductions to thifensulfuron treatment in WT were 91 to 99% for 0.5 to 32X the labeled rate while no reductions were observed in the suspected resistant populations up to 4X. From 8 to 32X, vigor reductions in the suspected resistant populations were 0 to 11%. Fresh weight reductions relative to the nontreated were 11 to 26%, 13 to 30%, 11 to 40%, and 20 to 55% for NK1, NK2, KW1, and KW2, respectively. These data clearly indicate resistance to thifensulfuron. The F2 generation of NK1 had fresh weight reductions of <23% up to 4X and <50% up to the 32X rate, indicating resistance heritability. Cross-resistance was observed to all herbicides tested in the NK1 population, which had vigor reductions of <10%. NK2 was not cross resistant, while both KW populations were only cross resistant to bispyribac. These data likely indicate different resistance-conferring mutations among the populations. Field data indicate that flumioxazin at â‰¥ 0.07 kg ai ha-1 applied PRE results in >95% common chickweed control. Delayed PRE applications of pendimethalin at 1.06 kg ai ha-1 and metribuzin at â‰¥ 0.05 kg ai ha-1 resulted in â‰¥ 94% control. POST applications including fluroxypyr at â‰¥ 0.12 kg ai ha-1 or metribuzin at â‰¥ 0.11 kg ai ha-1 resulted in â‰¥ 90% control, regardless of application timing (POST fall, POST early-spring, or POST late-spring).

MANAGEMENT OF ALS-RESISTANT WEEDS IN WINTER WHEAT. M.J. VanGessel*, Q.R. Johnson, and B.A. Scott, University of Delaware, Georgetown (56).

ABSTRACT

After years of relying exclusively on ALS-inhibiting herbicides for weed control in small grains, weed biotypes have developed resistance to this group of herbicides. Currently in DE and the mid-Atlantic region, common chickweed (*Stellaria media*), horseweed (*Conyza canadensis*), and annual ryegrass (*Lolium multiflorum*) have been increasing in small grains due to herbicide resistance. Various trials have been conducted to examine alternative herbicide treatments for management of these resistant biotypes. All trials were small plots procedures with backpack herbicide application. Treatments differed based on targeted weed species, but all trials included three replications.

Management of ALS-R horseweed was evaluated in one trial started in the fall of 2013 with Clarity, 2,4-D, Starane Ultra, and Huskie applied either in the fall or in the spring. Treatments containing Huskie and Clarity provided the highest level of control. Postemergence applications of Starane Ultra or 2,4-D did not provide acceptable control at commonly used rates.

Management of ALS-R annual ryegrass was evaluated in 2013 with fall or spring applications of Axial XL. Additional treatments included early postemergence applications of Axiom used in combination with Axial XL. Control of annual ryegrass with Axial XL was greater than 80% control, regardless of timing. There was a consistent trend with Axiom at planting followed by Axial XL improving full-season control by 5 to 10%, but this was not significant across all treatments with Axiom.

Four trials have examined control of ALS-R common chickweed. Across these trials, metribuzin was consistently one of the best treatments. Metribuzin applications in the fall or spring were effective, however, more winter wheat injury is observed with fall applications. Huskie also was also effective, but it was only included in two of the four trials. Starane Ultra provided good suppression of ALS-R common chickweed, however, it seldom provided over 75% control.

At this time, there appears to be effective options for the control of ALS-R biotypes in small grains. However, integrated approaches need to be developed to ensure that herbicide-resistant biotypes do not continue to develop or spread.

DUPONT AFFORIA HERBICIDE: NEW BURNDOWN OPTION WITH CROP ROTATION FLEXIBILITY. V. Kleczewski*, D.D. Ganske, K.A. Diedrick, and H.A. Flanigan, Dupont, Wilmington, DE (57).

ABSTRACT

Studies were conducted in the states of DE, MD, VA, NC, GA, KS, MS, and TX to evaluate pyroxasulfone applied PRE, delayed PRE (80% of germinated seedlings with shoots ½ inch in length), and POST (targeting 1 to 2 leaf ryegrass) alone and in mixtures for control of Italian ryegrass and other weed species. Pyroxasulfone (applied as Zidua® herbicide) was applied PRE and delayed PRE at rates between 0.5 to 1.25 oz/acre and was applied POST at rates between 1 to 1.5 oz/acre. POST mixtures of pyroxasulfone were evaluated alone and with pinoxaden (applied as Axial® herbicide) at 16.4 fl oz/acre and/or metribuzin (75% DF products) at 2 oz/acre. In general, PRE applications of pyroxasulfone caused more injury than delayed PRE applications, however, control of Italian ryegrass and jagged chickweed were slightly higher with PRE applications as compared to delayed PRE applications although both weeds were controlled 88% or greater regardless of these before emergence application timings. POST pyroxasulfone applied alone controlled Italian ryegrass, downy brome, and common chess 75% to 83%. In addition, pyroxasulfone POST did not control cereal rye and only provided 59% suppression of wild radish. The mixture of metribuzin plus pyroxasulfone applied POST increased wild radish and Italian ryegrass control to 99% and 86%, respectively. The most effective POST treatment was the three way mixture of pyroxasulfone plus metribuzin plus pinoxaden which provided 97 to 98% control of Italian ryegrass, common chess, and wild radish. The best pyroxasulfone treatments including the reduced rate PRE (0.5 to 0.67 oz/acre) fb a sequential POST (1 to 1.5 oz/acre) and the POST mixtures of pyroxasulfone with pinoxaden and/or metribuzin provided between 2 to 3 times greater yield when compared to the untreated check.

EVALUATION OF WEED CONTROL PROGRAMS UTILIZING HPPD-TOLERANT SOYBEANS. E.M. Hitchner*, D.E. Bruns, T.H. Beckett, B.R. Miller, and D.J. Porter, Syngenta Crop Protection, Elmer, NJ (59).

ABSTRACT

Field trials were conducted from 2012 to 2014 to evaluate mesotrione-based weed control programs in HPPD-tolerant soybeans stacked with glyphosate tolerance. These multiple mode-of-action herbicide tolerant soybeans enable the use of mesotrione and isoxaflutole pre-emergence in addition to glyphosate post-emergence.

Several mesotrione-based herbicide programs provided control of key weed species, including glyphosate resistant populations. The most successful and consistent weed control was achieved with two-pass programs that included pre-emergence residual herbicides and multiple, overlapping modes of action. These programs were designed to align with HRAC principles of weed resistance management. The use of these chemically diverse and novel programs will offer effective, safe and sustainable weed management options for soybean growers.

ACURON™: PREEMERGENCE WEED CONTROL AND CORN SAFETY. M.A. Cutulle*, T.H. Beckett, S.E. Cully, R.D. Lins, and G.D. Vail, Syngenta Crop Protection, Vero Beach, FL (61).

ABSTRACT

Acuron™ is a multiple mode-of-action herbicide premix that provides preemergence and postemergence grass and broadleaf weed control in field corn (as well as seed corn, sweet corn and popcorn). In addition to mesotrione, s-metolachlor, and atrazine, Acuron™ also contains bicyclopyrone, a new HPPD (4-hydroxyphenyl-pyruvate dioxygenase) inhibitor. Acuron™ applied preemergence is effective on difficult-to-control weeds, including common lambsquarters (Chenopodium album), common ragweed (Ambrosia artemisiifolia), giant foxtail (Setaria faberi), giant ragweed (Ambrosia trifida), Palmer amaranth (Amaranthus palmeri) and waterhemp (Amaranthus rudis) with improved residual control and consistency compared to commercial standards. Additionally, preemergence applications of Acuron™ are safe to corn. Pending regulatory approvals, first commercial applications are anticipated in the 2015 growing season.

CONTROL OF PALMER AMARANTH IN SOYBEANS. M.J. VanGessel*, Q.R. Johnson, and B.A. Scott, University of Delaware, Georgetown (62).

ABSTRACT

Palmer amaranth (*Amaranthus palmeri*) has been spreading in Delaware in the past few seasons. During the fall of 2013, over twenty-five soybean fields were infested with this species. Greenhouse trials confirmed that most of these uncontrolled Palmer amaranth plants were glyphosate-resistant. In addition, over 25% of the samples were resistant to Group 2 herbicides. Glyphosate-resistant Palmer amaranth has also been confirmed in Virginia, New Jersey, Maryland, and Pennsylvania.

A field trial evaluated the effectiveness of various approaches for Palmer amaranth control with genetically altered soybeans, utilizing Roundup Ready, sulfonylurea tolerant (STS), and glufosinate-resistant (Liberty Link) soybeans. Treatments consisted of a factorial arrangement of soybean genetics for herbicide resistance (Roundup Ready, Liberty Link, STS, and conventional soybeans) and four approaches to weed management (total postemergence [POST], preemergence followed by POST [28 days after planting], preemergence followed by late-POST [42 DAP], reduced rate of preemergence followed by POST). In addition, preemergence treatment followed by a higher rate of Liberty 280 was included as well as an untreated check for comparison. All treatments were replicated three times. The trial was conducted over four field seasons.

The locations were naturally infested with Palmer amaranth. The Palmer amaranth population was susceptible to both glyphosate and ALS herbicides. Across the four site-years, residual herbicides improved full-season Palmer amaranth control. At mid-season, Palmer amaranth control with only POST applications was rated 90% for glyphosate, but was poor for all other systems. High level of control for the glyphosate treatment was due to the presence of glyphosate-susceptible biotype. A program of PRE followed by POST was excellent for all four systems, but a reduced rate PRE fb POST was not as consistent across all site-years. A preemergence application followed by late POST treatment was often effective for Liberty-Link, Roundup Ready, and STS soybeans, but was not acceptable for conventional soybeans.

Management of Palmer amaranth requires a soil-applied followed by an effective postemergence herbicide. As resistant biotypes of Palmer amaranth continues to spread in the region, farmers need to select herbicides that will be effective for these biotypes and apply them when the Palmer amaranth plants are at susceptible stages of growth.

PALMER AMARANTH AND WATERHEMP: THE PIGWEEDS ARE SQUEALING IN PENNSYLVANIA. D. Lingenfelter* and W. Curran, Pennsylvania State University, University Park (63).

ABSTRACT

Palmer amaranth (*Amaranthus palmeri*), also known as Palmer pigweed, was recently introduced into Pennsylvania agricultural systems. It was identified on at least ten farms in 2013 and another 14 in 2014; and there are likely more. Infestations were primarily observed in soybean, corn, alfalfa, and around barnyards. It poses unique management challenges that most farmers in this region have not had to contend with before. Containing new infestations and preventing its spread is a critical first step to managing this new threat. The risk from this new weed comes from its competitive and aggressive growth habit, prolific seed production (greater than 100,000 seeds per plant), prolonged germination pattern, along with its potential resistance to glyphosate (Group 9) and the Group 2 herbicides (ALS-inhibitors). Resistance to other herbicide classes has also been documented in this species (Groups 3 and 5). Palmer amaranth is a summer annual broadleaf weed that is native to the southwestern US and Mexico. Palmer amaranth is related to other pigweeds in the Northeast region including redroot, smooth, Powell, and spiny, but unlike these other pigweeds, Palmer amaranth grows faster and is dioecious. Pollen from male plants can travel with the wind to susceptible female plants and if the male is herbicide resistant, a portion of the offspring will also be resistant. Waterhemp (*Amaranthus tuberculatus*), another dioecious herbicide resistant pigweed species common in the Midwest is also getting a foothold in Pennsylvania. Currently, there are at least three documented infestations and a few other sites have been reported. Although not as great a competitive threat as Palmer amaranth, it too should be aggressively managed to prevent its spread since it will certainly compete in Pennsylvania cropping systems. It is uncertain how Palmer amaranth and waterhemp populations were introduced to the farms, but equipment, hay, feed, manure, and compost are all suspected. Contaminated cotton seed from the south or west, used in dairy rations, may also be involved. Herbicide options for Palmer pigweed and waterhemp control can be limited in soybean, alfalfa, and vegetable productions systems; however more options exist in corn and pasture. Timely applications and a combination of preemergence/residual and postemergence herbicides containing at least two effective modes of action are critical for practical management of these species. Other management considerations include the use of incoming herbicide resistant crop varieties (i.e, 2,4-D-, dicamba-, and HPPD-resistant crop traits), more diverse crop rotations, cover crops, and mechanical tactics, among others. Education of farmers and agricultural service providers about these unwelcomed weed species, their negative impacts, and management are of vital importance.

ARE THERE NEW HERBICIDES WE CAN USE TO SUPPRESS PALE SWALLOW-WORT? A. Gover* and T. Rung, Penn State, University Park (65).

ABSTRACT

Pale swallow-wort (*Vincetoxicum rossicum* (Kleopow) Barbar.) is an exotic, herbaceous, perennial vine in the Apocynaceae family, native to the Black Sea region in Europe. It is currently reported in Quebec, Ontario, and in discrete locales in many northeastern and Great Lakes states. Reported empirical results to date indicate the utility of repeated applications of glyphosate, triclopyr, or imazapyr. We established a field trial to compare aminocyclopyrachlor-based products and tank mixes with glyphosate or triclopyr. The site was a cut slope along SR 34 near Carlisle, PA, planted to crownvetch (*Securigera varia* (L.) Lassen). Herbicides were applied July 2, 2013 to 3 by 5 m plots arranged in a randomized complete block with three replications, using a CO_2-powered handgun equipped with a TeeJet #5500 Adjustable Cone with an X-8 tip, targeting 190 L/ha. Treatments included a control; glyphosate at 3.4 kg/ha; triclopyr at 3.4 kg/ha; glyphosate plus triclopyr at 3.4 plus 1.7 kg/ha respectively; the premixtures of aminopyralid plus metsulfuron methyl at 0.12 plus 0.021 kg/ha, respectively, or aminocyclopyrachlor plus metsulfuron at 0.13 plus 0.042 kg/ha respectively, each with or without triclopyr at 1.7 kg/ha; and the combination of triclopyr plus the premixture of aminocyclopyrachlor plus metsulfuron plus imazapyr at 1.7 plus 0.066 plus 0.021 plus 0.091 kg/ha, respectively. Visual ratings of percent foliar necrosis were taken August 14, 2013 (6 weeks after initial treatment [WAIT]), and percent vegetative cover and percent swallow-wort cover was taken September 24, 2014 (68 WAIT). Each plot was retreated with the initial herbicide mixture September 17, 2013 (15 WAIT), targeting any green tissue. A Department of Transportation herbicide contractor mistakenly oversprayed the entire study area with triclopyr plus metsulfuron at 1.7 plus 0.021 kg/ha, respectively, on August 25, 2014. The data were subjected to analysis of variance, and means of significant effects were compared using a Protected L.S.D. When the untreated control was withheld from the analysis, there was not a significant treatment effect on swallow-wort necrosis, which averaged 42 percent. By contrast, crownvetch topgrowth was completely eliminated by all treatments. Swallow-wort cover at 68 WAIT was used to derive percent reduction (set to zero if negative). Perhaps most notable is that when the overspray is considered, the herbicide plots were treated three times in 13 months, and only 5 of the 27 plots were swallow-wort-free. The overspray-only ('control') plots averaged 20 percent reduction. The three treatments including aminocyclopyrachlor averaged 97 to 99 percent reduction, but were only significantly different from the control, triclopyr alone (62 percent), and glyphosate plus triclopyr (56 percent). This experiment was a somewhat coarse evaluation, but demonstrates that aminocyclopyrachlor-based treatments should be considered where site conditions permit.

CONVERSION OF CANADA THISTLE-INFESTED CROWNVETCH GROUNDCOVER TO FINE FESCUE TURF: A SEEDING TIMING EXPERIMENT. J. Johnson*, D.A. Despot, and J.C. Sellmer, Penn State University, University Park (66).

ABSTRACT

Crownvetch (*Coronilla varia*) is an effective groundcover on steep slopes with rocky mineral soils; however, it can be invasive and weedy where it develops unintentionally and it becomes overrun with difficult to control broadleaf weeds such as Canada thistle (*Cirsium arvense*). Herbicides that will spare the crownvetch yet provide some control of broadleaf weeds are limited. An effective approach would be a conversion strategy in which herbicides to control both the Canada thistle and crownvetch are applied and followed by the establishment of a fine fescue turf. One issue in implementing a conversion strategy is that some effective broadleaf herbicides persist in the soil after application and may inhibit the germination of desirable turf seeds. A conversion experiment was established at two locations to determine the effectiveness of various herbicide treatments to control crownvetch and Canada thistle while defining the best timing for turfgrass seeding after treatment. In this experiment, herbicide treatments were applied in the fall of 2012 followed by fine fescue seeding in the same season or as a spring application. The treatments included: 1) 123.4 g aminopyralid/ha, 2) 3365 g glyphosate/ha + 210.3 g clopyralid/ha, 3) 269.5 g aminocyclopyrachlor/ha + 14 g metsulfuron/ha, 4) 134.7 g aminocyclopyrachlor/ha + 7 g metsulfuron/ha + 1683 g triclopyr/ha, 5) a cut and seed treatment with no herbicide applied, and 6) a control treatment where no herbicides or seeds were applied. The plots were prepared by loosening the soil with a disc harrow immediately prior to seeding to assure good seed-soil contact. Plots were seeded with 7.06 kg hard fescue (*Festuca longifolia*)/1000 sq m + 4.62 kg creeping red fescue (*Festuca rubra*)/1000 sq m + 1.36 kg annual ryegrass (*Lolium multiflorum*)/1000 sq m and fertilized according to soil test recommendations. Two years after initiation of the experiment there was tremendous variability between the sites. At location one, the fall seeded plots averaged 68 percent fine fescue turf cover, whereas spring-seeded plots averaged 12 percent cover by fine fescue, whether herbicides were applied or not. We speculate that external factors reduced turf establishment in the spring-seeded plots (i.e., high vole population and competing perennial grasses). All treatment plots (fall and spring) resulted in an 80 to 97 percent reduction in Canada thistle except for two spring-seeded treatments which received 3365 g glyphosate/ha + 210.3 g clopyralid/ha and a cut only treatment each of which reduced thistle by 56 and 57 percent, respectively. The second location produced 39 to 80 percent turf cover across seeding times with two exceptions. First, a fall-seeded treatment that received 269.5 g aminocyclopyrachlor/ha + 14 g metsulfuron/ha and resulted in 21 percent turf cover. We speculate that the herbicide treatment may have inhibited turf establishment due to a four-week treatment to seeding window. Second a spring-seeded treatment where no herbicide was applied and 30 percent turf cover was recorded. At location two, all seeded treatments resulted in an 88 to 99 percent reduction in Canada thistle. Significant reductions in Canada thistle were observed at both sites. Employing herbicide treatments followed by seeding to fine fescue in either fall or spring also appears effective; however, site conditions may play a role in conversion success. This research represents preliminary work to develop a conversion and reseeding strategy for use along roadside corridors.

EFFECT OF PRE- OR POST-PLANT APPLICATIONS OF SULFOMETURON OR IMAZAMOX ON WOODY SEEDLINGS. T. Rung* and A. Gover, Penn State University, University Park (67).

ABSTRACT

A tank mix of glyphosate plus sulfometuron provides utility in riparian forest buffer plantings, particularly where reed canarygrass (*Phalaris arundinacea* L.) is prevalent. However, sulfometuron lacks aquatic labeling, which can restrict its use within riparian tree plantings. Imazamox could be a suitable alternative for sulfometuron if it provides similar vegetation suppression and does not damage transplants. Glyphosate at 3.4 kg ae/ha, alone or in combination with imazamox at 0.28 or 0.56 kg ae/ha, or sulfometuron at 0.053 kg/ha was applied pre- or post-planting to 6 by 6 ft sites around tree shelters planted to 1-year seedlings of arrowwood viburnum (*Viburnum dentatum* L.), gray dogwood (*Cornus racemosa* Lam.) or silky dogwood (*C. amomum* Mill.) arranged in a completely randomized design with a factorial treatment arrangement and four replications, on May 27, 2014 in a floodplain at Canoe Creek State Park, Hollidaysburg, PA. Tree mortality, percent vegetative cover and shoot growth data was collected on October 16, October 20, and November 5, 2014 respectively. Data were subjected to analysis of variance, and means of significant effects were compared using Fisher's Protected LSD.

No treatment had a significant effect on tree mortality, which averaged 14 percent across the entire study. Herbicide treatment was the only significant effect for vegetative cover and shoot growth. The most common species were common teasel (*Dipsacus fullonum* L.), crownvetch (*Securigera varia* (L.) Lassen), purple loosestrife (*Lythrum salicaria* L.), and goldenrod species (*Solidago* spp.). Glyphosate alone-treated sites had significantly higher vegetation percent cover (42 percent) than all other herbicides. Cover for the low rate of imazamox (25 percent) was not different from the high rate (19 percent), but was significantly greater than for sulfometuron (9 percent). Glyphosate-treated sites had the greatest amount of shoot growth (13.1 cm), which was significantly higher than the imazamox-treated shrubs (4.5 and 5.9 cm for the low and high rates, respectively) but was not significantly different from sulfometuron treatments (8.5 cm).

Another study initiated May 30, 2014 at the same site examined the same herbicide treatments for their suppression of a monotypic stand of reed canarygrass. When rated October 23, untreated plots had 100 percent reed canarygrass cover, while glyphosate alone plots averaged 21 percent, imazamox low and high rates averaged 11 and 10 percent, respectively, and sulfometuron-treated plots averaged 1 percent cover. These results suggest imazamox does not present a useful alternative to sulfometuron for weed control around sheltered woody seedlings due to inhibition of woody plant growth and less suppression of reed canarygrass compared to sulfometuron.

EFFECTIVENESS OF LONG TERM MONOECIOUS HYDRILLA MANAGEMENT PLANS IN NORTH CAROLINA. J.J. Nawrocki and R.J. Richardson*, North Carolina State University, Raleigh (68).

ABSTRACT

Hydrilla [*Hydrilla verticillata* (L.f.) Royle] is the most economically damaging aquatic weed in the United States. Long term hydrilla control is complicated by persistent subterranean turions (tubers) that the plant forms annually. Elimination of the tuber bank is essential for long term control or eradication efforts. Research was conducted on four North Carolina lakes to evaluate monoecious hydrilla tuber dynamics and to determine the effects of specific management techniques on monoecious hydrilla tuber numbers over time. Lake Gaston, Lake Tillery, Shearon Harris Lake, and the Tar River Reservoir were sampled for up to 7 years. Management practices and their effects on tuber density were assessed on each lake. Chemical control sites using fluridone were assessed on Lakes Tillery and Gaston whereas a combination of fluridone use, biological control through sterile grass carp, and physical control through drought induced summer drawdown was assessed on the Tar River Reservoir. Sites on Lake Gaston and Shearon Harris Reservoir with no active management were used as a control. De-watering and fluridone application in 2007 thru 2012 as well as a low density of grass carp stocking in 2013 resulted in an overall decrease in tuber density of 100% in the Tar River Reservoir. Two tubers found on the Tar River Reservoir in fall 2012 were assumed to be 6 years or older and were still viable. Lake Gaston sites subjected to fluridone treatment every other year demonstrated a tuber bank reduction of 26% after 2 years and 60% after 4 years. Sites on Lake Gaston that were treated consecutively for 2 years exhibited a 75% reduction in tuber density. On the unmanaged Shearon Harris Reservoir, average whole lake densities ranged from 838 to 2,050 tubers per m2 from 2008 to 2013. At a single sample site a density of 3,244 tubers was recorded in the fall of 2008, which is higher than previously reported in situ.

EVALUATION OF HERBICIDE MIXTURES FOR CONTROL OF MORROW'S HONEYSUCKLE. J. Johnson*, D.A. Despot, and J.C. Sellmer, Penn State University, University Park (69).

ABSTRACT

The proliferation of exotic shrub honeysuckle species including tatarian (*Lonicera tatarica*), amur (*Lonicera maackii*), and Morrow's (*Lonicera morrowii*) honeysuckle has become a common problem along the roads of Pennsylvania. Once established these woody plants create limited sight distance and interfere with the maintenance activities on the roadside. Herbicide mixes used to target other brush species are often ineffective on exotic shrub honeysuckles. Efforts to identify herbicide mixes in previous experiments have shown that glyphosate and 2,4-D are generally effective; however, each presents problems. Glyphosate is not selective and damages the grass understory. Conversely, 2,4-D does not damage the grass understory but honeysuckle resprouting can occur. The goal of this experiment was to build upon previous efforts and determine if increased rates of 2,4-D in tank mix combinations will offer greater control. Eight treatments including seven herbicide combinations were tested including: 269.5 g aminocyclopyrachlor and 13.9 g metsulfuron plus 3197 g 2,4-D acid/ha; 42.1 g metsulfuron plus 4263 g 2,4-D acid/ha; 1683 g triclopyr plus 2131 g 2,4-D acid/ha; 1683 g triclopyr plus 4263 g 2,4-D acid/ha; 1683 g triclopyr plus 3365 g glyphosate acid/ha; 2356 g 2,4-D acid and 897.5 g triclopyr/ha plus 31.6 g metsulfuron/ha; 4483 g 2,4-D acid and 1725.5 g triclopyr/ha plus 21.1 g metsulfuron/ha; 3365 g glyphosate acid/ha; and an untreated check. All herbicides treatments included a non-ionic surfactant at 0.25 percent v/v.

The shrubs targeted in this experiment were on average 3 meters in height. As a result, using our standard CO_2 plot sprayer, the foliar spray pattern was limited to side trimming many of the plants within the treated area and later assessing control of branches located within the application zone. Treatments containing glyphosate offered the greatest control of Morrow's honeysuckle nearly one year after treatment, YAT, with values of 94 to 97 percent control. Moderate control was achieved with 2,4-D acid at rates of 3197 g/ha or higher with control values of 69 to 75 percent. The remaining treatments resulted in lower control ratings from 31 to 54 percent. Injury to grasses was observed with treatments containing glyphosate. Further testing of these herbicide combinations including complete coverage sprays are necessary to assure efficacy in control of the troublesome exotic honeysuckles along with other woody species prior to making a clear recommendation for their use.

JAPANESE KNOTWEED'S PROGENY: A REVIEW OF ITS HYBRID FROM MANY SOURCES. T.W. Curran*, U.S. Army Corps of Engineers (retired), Huntington, WV (72).

ABSTRACT

The knotweed complex as typified by *Polygonum cuspidatum* Siebold & Zucc. consists of, as reclassified into Fallopia, *F. japonica* (Houtt.) Ronse Decraene (Japanese knotweed), *F. sachalinensis* (F. Schmidt) Ronse Decraene (giant knotweed) and their hybrid *F. x bohemica* (Chrtek & Chrtkova) J.P. Bailey (Bohemian or hybrid knotweed). The hybrid is accepted as the most invasive of the three, and the most difficult to control. Although the purpose of this review is to focus on the hybrid, because of close similarities among the three, the review must also include the parents of the hybrid. All three species have been recorded as causing infrastructure damage, such as breaking up asphaltic paving, and heavily damaging masonry walls. Because of behavior that could be considered unusual, idiosyncratic, and possibly even bizarre, these three species of noxious weeds, especially the hybrid, present many difficulties to the researcher and to the land manager. Although the knotweed complex has been the subject of extensive and varied peer-reviewed research, additional research areas for research have become apparent when available information about the complex is considered in concert with published knowledge of physiological processes in common with or similar to those of other plants. This brief synoptic review taken from many credible sources shows possible scientific evidence to explain such behavior as the rhizome's observed revival from dormancy after burial for many years, and why cutting the plant to the ground early in the growing season could give the plant an advantage to survive herbicide application in the fall.

The robustness of the hybrid's rhizome as first measured by Petr Pysek was found to surpass that of both parents. However, even parental rhizome is known to have survived burial under up to three feet of volcanic ash. Regrowth of knotweed treated with herbicide has been observed to produce epinastic leaves. Epinasty in large plants is associated with anoxic conditions caused by submergence within the root zone. Anoxic metabolism, or lactate metabolism, in cellular respiration is more conservative of plant carbohydrate energy stores than aerobic metabolism, but causes plant tissue to be subject to a greater degree of oxidative stress. Plants tolerant of anoxic tissue conditions have higher levels of antioxidant compounds than those not as tolerant. Therefore, it is not surprising that Japanese and giant knotweed rhizome supplies the raw material for the majority of the global market for food supplements of resveratrol, a powerful antioxidant. The long-term survival of knotweed rhizome was a normal part of the plant's adaptation to its environment, and is not abnormal or bizarre.

Similar investigation shows that cutting of knotweed early in the growing season, to weaken the plant, might be counter-productive to overall control measures. Also, considering ploidy of Korean knotweed specimens, variable ploidy within the hybrid may not be unusual.

SO MANY WEEDS, SO LITTLE TIME: A SITE-SPECIFIC TOOL TO PRIORITIZE INVASIVE SPECIES SUPPRESSION. A. Gover* and R.K. Wagoner, Penn State, University Park (73).

ABSTRACT

In a given native plant community, invasive plant species are often the most detrimental element that can be influenced by management. Optimal use of finite human and material resources requires a triage process that accounts for the inherent values of a specific site and the effect of the invasive species in that site. The State Park system in Pennsylvania uses a five-factor index that accounts for the properties of the specific land sub-unit and the invasive species that are present or incipient. Prior to calculating the index, the property is delineated into units that have ecological and operational resonance. A suggested filter at this step is an initial habitat/non-habitat designation so that only areas with a threshold level of ecological value are considered. Then the units are assessed according to Stewardship Value, Invasive Extent, Invasive Impact, Restoration Effort, and Outreach Value. Values for Stewardship and Outreach are assigned to the land unit, and Extent, Impact, and Effort values are assigned for each species in each unit. Stewardship Value accounts for the ecological integrity of the units. Invasive species Extent characterizes each species as leading edge (high value), significant and still expanding, or saturated (low value) within the unit. A positive Impact value is assigned if that species is known to be well adapted to spreading in that unit. Restoration Effort characterizes the work trajectory for a given invasive species, accounting for its regenerative capacity. Outreach Value accounts for community interest or potential for external collaboration as added value. Additionally, a presence/absence value is assigned for each cell. The system calculates a score of 0 to 10 for each cell in the species-by-site matrix, adding Stewardship Value (0-2), Invasive Species Extent (0-2), Invasive Species Impact (0-1), Restoration Effort (0-3), and Outreach Value (0-2), and then multiplying by presence/absence (0 or 1). Prescriptions are then developed and scheduled for the highest priority sites based on seasonal availability of resources.

The effort required to populate all the cells in these matrices and implement the treatment is mitigated by the concept of 'bundling'. When an invasive species in a unit triggers a priority response, all practicable target species are treated. This has two significant implications. First, a limited number of treatments are developed that have broad applicability, so that almost any invasive species present during an operation is treated. Second, if one species is important enough to trigger a response, you do not need extensive inventory data of all species. This approach requires an emphasis on training staff to identify target species, which allows for a flexible, adaptive response once they are on-site, and accomplishes survey and monitoring during operations.

ANNUAL BLUEGRASS SEEDHEAD SUPPRESSION IN SIMULATED FAIRWAYS. J.A. Borger* and T.L. Harpster, The Pennsylvania State University, University Park (75).

ABSTRACT

This study was conducted over two years on a mature mixed sward of annual bluegrass (*Poa annua*) and creeping bentgrass (*Agrostis stoloifera)* at the Valentine Turfgrass Research Center, Penn State University, University Park, PA. The objective of the study was to determine if selected materials applied in the spring could suppress annual bluegrass seedhead populations under simulated golf course fairway conditions.

The studies were randomized complete block designs with three replications. Treatments were applied on 15 April, 3 May and 17 May 2013, and on 4 April, 12 April and 1 May, 2014 using a three foot CO_2 powered boom sprayer (Bellspray, Inc., Opelousas, LA) calibrated to deliver 40 gpa using one, flat fan, TP9504EVS nozzle (TeeJet Technologies, Wheaton, IL) at 50 psi.

The test sites consisted of approximately 95 percent annual bluegrass and 5 percent creeping bentgrass at the initiation of the study. Turfgrass populations were visually evaluated for the percent seedhead coverage in order to evaluate the test material's ability to suppress annual bluegrass seedheads. Additionally, turfgrass was irrigated on an as needed basis to prevent moisture stress.

Data was analyzed with ARM 8.5.0 (Gylling Data Management, Inc., Brookings, South Dakota) using Duncan's New MRT at the 0.5 percent significant level.

The addition of Civitas and Harmonizer to Embark reduced the seedheads (some dates significantly) when compared to turfgrass treated with Embark alone. Also the combination of these products revealed that when a lower rate of Embark was employed (24 oz vs 48 oz) seedhead suppression was achieved.

In conclusion, it is apparent that the addition of Civitas and Harmonizer made a significant difference in the PGRs efficacy in these studies under these conditions in the suppression of annual bluegrass seedheads.

SELECTIVITY OF PRIMISULFURON-METHYL FOR ANNUAL BLUEGRASS CONTROL IN KENTUCKY BLUEGRASS. P. McCullough*, M. Czarnota, C. Johnston, and R. Singh, University of Georgia, Griffin (76).

ABSTRACT

Annual bluegrass (*Poa annua*) is a problematic weed of Kentucky bluegrass (*Poa pratensis*) turf that may be selectively controlled with POST applications of primisulfuron-methyl. The objective of this research was to evaluate physiological behavior of primisulfuron-methyl attributed to selectivity in these species. In application placement experiments, annual bluegrass shoot weight reductions from the nontreated from high to low were treatments including: foliar + soil ≥ soil only ≥ foliar only. Annual bluegrass averaged 33 and 52 percent shoot weight reductions from the nontreated after 4 wk from primisulfuron-methyl at 40 and 80 g ha^{-1}, respectively. Kentucky bluegrass shoot weight was not reduced from the nontreated and application placements were similar. From five harvests ranging 1 to 168 hours after treatment (HAT), annual and Kentucky bluegrass absorbed up to 25 and 32 percent of foliar applied ^{14}C-primisulfuron-methyl, respectively, and reached peak absorption at 22 (±1.4) and 31 (±1.6) HAT, respectively. Both grasses distributed 15% of foliar absorbed ^{14}C to nontreated shoots with minimal translocation (≤2 percent) to roots after 168 h. Annual bluegrass translocated 2x more root-absorbed ^{14}C to shoots than Kentucky bluegrass at 24, 72, and 168 HAT. From foliar uptake, metabolism increased with time from 1 to 168 HAT but Kentucky bluegrass averaged ≈50 percent less parent herbicide than annual bluegrass and 27 percent greater levels of a polar metabolite (R_f 0.1). In root metabolism experiments, annual bluegrass had ≈3x and 2x more primisulfuron acid in roots and shoots, respectively, than Kentucky bluegrass at 24, 72, and 168 HAT. The isolated ALS enzymes from the two grasses were equally susceptible to inhibition by primisulfuron-methyl. Overall, selectivity of primisulfuron-methyl for annual bluegrass control in Kentucky bluegrass is attributed to differential translocation and metabolism between species.

COMPARING HERBICIDES FOR THE CONTROL OF FALSE GREEN KYLLINGA. C. Mansue* and J.A. Murphy, Rutgers University, New Brunswick, NJ (77).

ABSTRACT

During the past two years, turf managers in southern New Jersey have reported difficulty controlling false green kyllinga (*Kyllinga gracillima*). Greenhouse and field trials were conducted to assess the efficacy of herbicides for controlling false green kyllinga during 2014. The greenhouse trial was conducted on plants collected from Atlantic County, NJ and the field trial was conducted on a soccer field in Ocean County, NJ. Field plots on the soccer field were comprised of 58 to 82% false green kyllinga. Both trials evaluated sulfentrazone, halosulfuron and mesotrione at label rates and timings. In the greenhouse trial, sulfentrazone applied at 0.14 kg a.i. ha^{-1} resulted in 63% injury of false green kyllinga after one month and increased to 99% one month after a sequential application. Halosulfuron applied at 0.052 kg a.i. ha^{-1} resulted in 13% injury of false green kyllinga after one month; injury increased to 33% one month after a sequential application. Mesotrione applied at 0.175 kg a.i. ha^{-1} resulted in 20% injury of false green kyllinga after one month; 97% injury was observed one month after a sequential application. In the field trial, a single application of sulfentrazone 0.14 kg a.i. ha^{-1} resulted in only 10% injury of false green kyllinga and a sequential application increase injury to 42%. Conversely, halosulfuron applied at 0.052 kg a.i. ha^{-1} produced 40% injury of false green kyllinga after one month and 77% injury after a sequential application. Mesotrione applied at 0.175 kg a.i. ha^{-1} produced 8% injury of false green kyllinga and increased injury to 35% after a sequential application. Although none of the herbicides were 100% effective at controlling false green kyllinga in the field, halosulfuron was more effective at suppressing false green kyllinga than sulfentrazone and mesotrione.

EFFECT OF METHIOZOLIN ON 13C-GLUCOSE ASSIMILATION IN ANNUAL BLUEGRASS AND THREE TURFGRASSES. K.A. Venner*, S.D. Askew, E. Colla'kova', and S. Koo, Virginia Tech, Blacksburg (78).

ABSTRACT

Methiozolin (PoaCure) is a new herbicide developed by the Moghu Research Center in Daejeon, Korea for the safe and selective removal of annual bluegrass (*Poa annua)* from creeping bentgrass (*Agrostis stolonifera*) putting greens. Methiozolin is a member of the isoxazoline class of chemistry, and, although two modes of action have been proposed, its mode of action is unknown. In *Lemna paucicostata* and other broadleaf species, Grossmann et al. (2011) proposed that methiozolin inhibits the function of tyrosine aminotransferase, an important enzyme in the conversion of L-tyrosine to 4-hydroxyphenyl pyruvate (4-HPP). Ultimately, the disruption in this pathway prevents catalysis of phytoene desaturase, an important cofactor in the formation of plastoquinones and tocopherols, important compounds in the protection of plant tissues from oxidative stress. The other proposed mode of action is cell wall biosynthesis inhibition. Lee et al. (2007) proposed that methiozolin acts as a cell wall biosynthesis inhibitor (CBI) based on research suggesting this herbicide inhibits the incorporation of 14C-glucose into plant cell walls. In studies investigating the incorporation of 14C-glucose in corn root tips, Lee et al (2007) noted decreased levels of incorporation in the cellulose and hemicellulose constituents of the cell walls, similar to what is observed with known cell wall biosynthesis inhibiting herbicides, like dichlobenil. Again, similar to other CBI-type herbicides, methiozolin induced a change in the coleoptile of barnyardgrass seedlings, broadening and thinning the coleoptile. A color change was also noted. Although these symptoms are different from other known CBI-type herbicides, the morphological change could further indicate a CBI mode of action.

Studies were conducted at Virginia Tech in 2014 in order to elucidate whether or not methiozolin influences the incorporation of 13C-labeled glucose into the cell walls of annual bluegrass (*Poa annua*), Kentucky bluegrass (*Poa pratensis*), creeping bentgrass (*Agrostis stolonifera*) and perennial ryegrass (*Lolium perenne*), one susceptible and three known tolerant turfgrass species to methiozolin, respectively. The treatment design was a factorial, turfgrass species by herbicide rate. Treatments were as follows: methiozolin at 0.01 and 1.0 uM, methiozolin at 1 uM + 4-HPP, indaziflam at 700 pM and a non-treated was included for comparison. Plants were treated for 24 hours in herbicide solution and subsequently transferred to a 10mM solution of 13C-glucose for 72 hours to allow for incorporation. Hydrolized cell walls were derivitized to individual hexose and pentose sugars. Each sample was evaluated using a gas chromatography-mass spectrometer. Preliminary data analysis suggests that levels of 13C-glucose incorporation in cell walls was poor, but analysis is ongoing.

METHIOZOLIN SOIL SORPTION IN SAND-BASED ROOT-ZONES. M.L. Flessner*, S. McElroy, G. Wehtje, and J. Howe, Virginia Tech, Blacksburg (79).

ABSTRACT

Methiozolin controls annual bluegrass (*Poa annua*) in golf course putting greens. Previous research indicates root-exposure is necessary for maximum herbicide efficacy. Therefore, methiozolin must be available for root uptake in the soil for successful field application; methiozolin must not be excessively sorbed to soil and not readily leached from the root-zone. Research was conducted to evaluate methiozolin sorption to various sand-based soil media, typical of many golf course putting greens. Soil solution batch-equilibrium methods were used with ^{14}C-methiozolin at 1 mg L^{-1} to determine sorption (Kd) values for United States Golf Association (USGA) grade sand with 0 to 20% (w/w) peat, USGA grade sand-based root-zones that were established as putting greens for 0, 10, and >15 years, and a Wickham sandy loam (locally collected field soil). Results indicate that organic matter content greatly influenced soil sorption, while soil pH did not influence soil sorption. Clay content was not included in the analysis due to its narrow range (1.3 to 7.5%) in the soil media evaluated. Kd values ranged from 0.4 mg L^{-1} for pure sand to 29.4 mg L^{-1} for sand with 20% peat. Native field soil had a Kd value of 6.4 mg L^{-1}. Kd values observed are consistent with other soil-active herbicides. Soil media with 0.3 to 6.1% organic matter resulted 9.5 to 32.1% of applied methiozolin in solution after equilibrium. Averaged across soil media, 24% of applied methiozolin is available for root uptake. This level of sorption suggests resistance to loss through leaching displacement.

LAST CALL HERBICIDE: CLOSING TIME FOR CRABGRASS AND GOOSEGRASS.
J.M. Calabro*, J. Fausey, and R. Fletcher, Nufarm Americas, Annapolis, MD (80).

ABSTRACT

Last Call™ is a selective postemergence herbicide for weed management in cool-season turf and Zoysiagrass. This product is a patent pending emulsifiable concentrate formulation containing fenoxaprop, fluroxypyr and dicamba that provides strong control of crabgrass, particularly in the early and mid-tiller stage of growth, and goosegrass. University research studies have shown crabgrass control ratings up to 92% when applied at the 2-3 tiller stage, as compared with 72% control with quinclorac. Goosegrass control studies reported greater than 90% control following June applications. Last Call also controls a number of broadleaf weeds including dandelion, clover and lespedeza and has shown strong activity against common bermudagrass in zoysiagrass and cool-season turf.

NEW HERBICIDES FOR VIRGINIA BUTTONWEED CONTROL. K.M. Vollmer*, S.D. Askew, B. Walls, and J. Hope, Virginia Tech, Blacksburg (81).

ABSTRACT

Virginia buttonweed (*Diodia virginiana* L., VB) continues to be one of the most troublesome weeds of southern turfgrass. Postemergence control can be achieved with metsulfuron or metsulfuron mixed with hormone herbicides but VB often recovers via seed germination or sprouts from surviving stem material within a few months of treatment. Better solutions are needed for long-term VB control. Three herbicides marketed in the last five years; Blindside (metsulfuron + sulfentrazone; FMC Corp), Tribute Total [foramsulfuron + halosulfuron + thiencarbazone (FHT), Bayer], and Specticle (indaziflam, Bayer); may offer improved VB control. Our objectives were to evaluate VB control with rates of metsulfuron + sulfentrazone with and without surfactant compared to metsulfuron and various programs that included indaziflam alone or in combination with FHT at different application times. Two studies were conducted as randomized complete block designs with 3 replications per treatment. Applications were made using a 1.5 m boom with XR 110003 nozzles delivering a flow rate of 274 L ha^{-1}. Data were subjected to analysis of variance and means separated using Fisher's LSD ($\alpha = 0.05$). In the first study, treatments included indaziflam applied at 33 or 49 g ai ha^{-1} on June 9, 2014, indaziflam at 33 g ha^{-1} on June 9, 2014 followed by thiencarbazone-methyl (22 g ai ha^{-1}) + foramsulfuron (44 g ai ha^{-1}) + halosulfuron-methyl (69 g ai ha^{-1}) (FHT) on July 8, 2014, indaziflam applied on June 16, 2014 followed by FHT on August 8, 2014, and FHT alone applied on June 16, 2014.

Treatments in the second study consisted of sulfentrazone (273 g ai ha^{-1}) + metsulfuron (27 g ai ha^{-1}), and sulfentrazone (421 g ai ha^{-1}) + metsulfuron (42 g ai ha^{-1}) each applied with and without the addition of non-ionic surfactant NIS (0.25% v/v) and compared to metsulfuron applied at 42 g ai ha^{-1}. Indaziflam discolored VB up to 40% in the first 2 weeks but VB completely recovered. Indaziflam fb FHT at the early timings controlled VB 90% but VB completely recovered by October 9. Indaziflam fb FHT at the late timing controlled VB 97% and subsequent weather conditions prevented VB recovery. Metsulfuron + sulfentrazone with surfactant at either rate controlled VB up to 95% and superior to metsulfuron + sulfentrazone at the low rate without surfactant and metsulfuron. Metsulfuron combinations controlled VB for approximately 6 weeks but VB completely recovered from all treatments by October 9th in the metsulfuron study. These data suggest new herbicides may slightly improve VB control compared to standard programs but will not control VB season-long.

REDUCING VISUAL IMPACT OF TOPRAMEZONE ON CREEPING BENTGRASS FAIRWAYS USING TRICLOPYR ADMIXTURES. S.D. Askew*, Virginia Tech, Blacksburg (82).

ABSTRACT

Topramezone was registered for use in turfgrass as Pylex herbicide in 2013. Use of topramezone on creeping bentgrass has formerly been prohibited. Preliminary research at Virginia Tech, BASF, and other universities have suggested commercially-viable uses for topramezone may exist in creeping bentgrass turf. Topramezone can cause severe discoloration to creeping bentgrass but turf typically recovers in less than one week. By lowering the use rate to 6 g ai/ha, creeping bentgrass fairway discoloration can be minimized. This rate will control goosegrass when applied once but repeated treatments are needed to control crabgrass or bermudagrass. Even at the 6 g/ha rate, creeping bentgrass discoloration can exceed 50%. Transient creeping bentgrass injury can be tolerated if fairways are treated only once or twice during the season but more applications are needed for some targeted weeds. At Virginia Tech, several studies have indicated triclopyr can eliminate foliar discoloration caused by topramezone. Preliminary studies indicated triclopyr may cause unacceptable injury to creeping bentgrass when used at rates above 70 g ai/ha leading us to hypothesize that the optimal rate to reduce creeping bentgrass discoloration is less than 70 g/ha. Our objectives in 2014 were to evaluate how 35 g/ha triclopyr will influence creeping bentgrass response to topramezone at 6 and 12 g/ha and to demonstrate a broad range of triclopyr rates in mixture with topramezone. Two studies were conducted at the Glade Road Research Facility at Virginia Tech in Blacksburg. Both studies were arranged as randomized complete blocks with 3 replications. Topramezone was applied at 6 and 12 g/ha alone or mixed with 35 g/ha triclopyr. One study consisted of 'L93' creeping bentgrass infested with smooth crabgrass and the other was 'Midnight II' Kentucky bluegrass infested with goosegrass and smooth crabgrass. Both sites were maintained at 1.5 cm mowing height. Topramezone alone injured creeping bentgrass approximately 30 and 70% when applied at 6 and 12 g/ha, respectively. When 35 g/ha triclopyr was added to topramezone, creeping bentgrass was not injured. At the creeping bentgrass site, all treatments completely controlled smooth crabgrass after two treatments at 3-week intervals. Kentucky bluegrass was never injured by topramezone. Goosegrass (3-10 tiller) was controlled 100% by all treatments. Smooth crabgrass was controlled 60% by two treatments of 6 g/ha topramezone alone and 93 to 98% by 12 g/ha topramezone or 6 g/ha topramezone plus 35 g/ha triclopyr. A nonreplicated demonstration indicated that optimal triclopyr rate in mixture with topramezone is between 12 and 35 g/ha. In 2014, the Pylex label was expanded to include uses in creeping bentgrass fairways and these data should help prevent undesirable turfgrass discoloration in this new market.

MULTI-YEAR EVALUATION OF PREEMERGENCE HERBICIDES. S. McElroy*, M.L. Flessner, and P.F. Aldahir, Auburn University, Auburn, AL (84).

ABSTRACT

In data presentation, tables or bar charts are normally utilized to present the mean of a given treatment accompanied by mean separation procedures such as Fisher's Protected LSD. In such a comparison, researchers are able to statistically separate treatments based on the mean. While statistically sound, end-users are only given partial information in such a scenario to base their herbicide selection decisions. In turfgrass management, end-users are specifically interested in the consistency of products on a year after year basis, thus it is the treatment variability that is also of interest. Box-and-whisker plots, also referred to as box plots) are ideal for visualizing and comparing variability of treatments when trials are conducted over multiple years or locations. Research was initiated to evaluate preemergence weed control in turfgrass for four consecutive years with the goal of identifying possible yearly variability in treatments.

A preemergence herbicide evaluation trial was initiated in Fall 2009 to evaluate annual bluegrass control and Spring 2010 to evaluate smooth crabgrass control. Each trial was repeated yearly in a different location until 2014, with trial initiation occurring in early March for preemergence smooth crabgrass control (four trials total) and mid-September for annual bluegrass control (four trials total). Treatments included (in kg ai/ha): atrazine (1.12), prodiamine (1.12), prodiamine (0.76) plus sulfentrazone (0.36) dithiopyr (0.56), pendimethalin (2.2), pendimethalin (2.2) plus dimethenamid-p (1.68), simazine (1.12), oxadiazon (3.36), indaziflam (0.027), indaziflam (0.054), and dimethenamid-p (1.68). Data were rated for percent weed control on a 0 to 100% scale monthly, however we present data from the conclusion of experiments which was early April for annual bluegrass trials and late-August for smooth crabgrass control trials. Treatments were applied with a CO_2 pressurized backpack sprayer at 280 L ha^{-1}. No adjuvants were included with any treatment. Data subjected to analysis of variance using PROC GLIMMIX using SAS 9.2 (SAS Institute, Cary NC). Box and whisker plots were generated using PROC BOXPLOT.

First in evaluating annual bluegrass control, only prodiamine and indaziflam at 0.046 lower quartile range remained above 80% control. Quartile range of oxadiazon and indaziflam at 0.027 was 60 to 85% and 76 to 95%, respectively. For smooth crabgrass, only indaziflam (0.054) lower quartile exceeded 90%. Minimum lower quartiles for pendimethalin, pendimethalin plus dimethenamid-p, prodiamine plus sulfentrazone, dithyopyr, and prodiamine was 70% control or greater. Lower quartiles for atrazine, oxidiazon and indaziflam (0.027), failed to reach 60% control. Based on this results, prodiamine and indaziflam (0.054) can used to consistently and efficiently control annual bluegrass yearly. For smooth crabgrass, efficient and consistent control was only achieved with indaziflam (0.046), yet several herbicides resulted in moderate-to-high control.

ASPECTS OF INDAZIFLAM PERFORMANCE FOR GREENHOUSE APPLICATIONS.
A. Senesac*, Cornell Cooperative Extension, Riverhead, NY (85).

ABSTRACT

With the recent EPA approval of indaziflam (Marengo) for application in enclosed greenhouses and hoop houses, there are several issues that need to be addressed concerning the application and efficient use of this preemergent herbicide. One such question concerns application along the edges inside empty houses that will soon be filled with potentially sensitive bedding plants. If indaziflam sprays are applied to plastic or glass sides, will there be an injurious 'splash back' occurring later when regular irrigation resumes once the house is re-occupied? A study was conducted at the LIHREC in 2014 to answer this question. An empty unheated hoop house that was covered in white polyethylene film with a woven plastic landscape fabric ground cover was the substrate for the treatments. The sprays were applied in one of three separate zones: Zone1-a two foot swath along the floor from the edge of the vertical rise. Zone 2- a two foot swath applied along the vertical wall starting where the wall and floor intersect. Zone 3- a 2 foot swath applied four feet up on the vertical curve of the hoop house wall. These areas were used to simulate possible locations of intentional or inadvertent spray deposit. Following treatment, in order to realistically simulate exposure, flats of six commonly grown ornamental and vegetable transplants were placed in three locations near and beneath the spray zones. The treatments consisted of 14.5 fluid oz. per acre of Marengo 0.622 SC (indaziflam) applied to either of the zones as well as a treatment that included all three zones. This 'thee zone' treatment was compare to Sureguard 51 WDG (flumioxazin). The flats were moved into position eight hours following the spray treatments. The flats were irrigated daily by hand for the next 21 days. After 21 days, the trial plants were harvested and aboveground fresh weight, as well as visual ratings of plant responses, was measured.

The results indicate that of the transplant species tested (coleus, petunia, portulaca, parsley and two tomato cultivars), coleus and portulaca were the most sensitive to indaziflam. Even so, the injury level did not rise above 18% for any species exposed to indaziflam. Flumioxazin, which is not labeled for use in closed structures, caused much greater injury to all six plant species. The results indicate that there is a fairly high level of tolerance of bedding plants to inadvertent indaziflam exposure in a closed greenhouse.

EVALUATION OF TWO FORMULATIONS OF INDAZIFLAM FOR ORNAMENTAL WEED CONTROL. H.M. Mathers* and L.T. Case, Ohio State University, Columbus (86)

ABSTRACT

Preemergence herbicides are of the utmost importance to the nursery and landscape industries. Indaziflam is a new preemergence herbicide developed by Bayer Corp. that has great potential for nursery and landscape use. However, the formulation of the active ingredient can affect herbicide activity, increase worker safety, and have effects on the environment. OHP has developed a "verge" formulation that decreases dust, is easier to spread, and spreads more evenly that the original "eco" formulation developed by Bayer. The "verge" technology has been applied to Biathlon (oxyfluorfen + prodiamine) herbicide, also distributed by OHP. The objective of this study was to evaluate any differences in weed control of four common nursery weeds between the "verge" formulation and the "eco" formulation. Four weed species were selected to determine efficacy, which included yellow foxtail (*Setaria gluaca*), large crabgrass (*Digitaria sanguinalis*), spiny sowthistle (*Sonchus asper*), and common yellow woodsorrel (*Oxalis stricta*). On 7 July 2014, three rates of indaziflam "verge" or "eco" were then applied over the top of the containers using shaker jars; treatments consisted of each formulation of 100 lbs/ac, 200 lbs/ac, or 400 lbs/ac. Evaluations consisted of weed counts at 1 WAT (week after treatment), 2 WAT, 4 WAT, 8 WAT, and 12 WAT and a weed fresh weight at 8 and 12 WAT. Both formulations at all rates provided excellent control until 8 WAT. At 12 WAT, there was some separation between the 100 lb/ac rate and 400 lb/ac rate of both formulations with the yellow foxtail, common yellow woodsorrel and spiny sowthistle. The large crabgrass continued to be controlled by all rates of both formulations. This data also supports the "verge" technology in that there are no differences between the two formulations, which indicates that the "verge" technology can be utilized for the indaziflam active ingredient.

COMPARISON OF CONVENTIONAL AND NEW HERBICIDES FOR ROBUST NURSERY FIELD WEED CONTROL. H.M. Mathers* and L.T. Case, Ohio State University, Columbus (87).

ABSTRACT

Many nursery field weed control programs pre-2008 relied heavily on manual methods of control. Between 2009 and 2013, extensive cuts were made in weed control budgets as the nursery market suffered under the downturned economy. Luckily, many new herbicides were researched and released during the 2008-2013 period. The objective of this research was to determine if some of these new versus conventional preemergence herbicides were capable of controlling the aftermath, exaggerated weed pressures of these downturned years. Trials were conducted in *Buxus* 'Green velvet' and *Taxus densiformus* fields at Studebaker Nurseries, New Carlisle, OH in 2013. Products tested were V-10366 (flumioxazin + pyroxasulfone, Valent U.S.A.) at 7.5, 15, and 30 oz/ac; Tower + Pendulum AquaCap (dimethenamid-p + pendulum, both from BASF Corp.) at 32 oz/ac + 2 qt/ac, respectively; and SureGuard (flumioxazin, Valent U.S.A.) at 6 oz and 12 oz/ac. Treatments were applied May 6 and reapplied on June 17, 2013. Liquid applications were applied as directed sprays. For both boxwood and yew, there were four replications/treatment and three subsamples/replicate, arranged in a completely randomized design within rows of each species. Treatments were evaluated at 1 WAT (weeks after treatment) 2 WAT, 4 WAT and 1 WA2T (weeks after second treatment). All treatments provided commercially acceptable weed control ($\geq$ 7) through 4 weeks after application. Only the V-10366 at 30 oz/ac was commercially acceptable 1 WA2T. By the second application, there was severe perennial weed pressure including Canada thistle, field bindweed, dandelion and wild garlic that only the V-10366 at the high rate could control. All products caused little phytotoxicity, well below commercially acceptable, at all evaluation dates.

MANAGEMENT OF WEEDY SPRING BULB SPECIES. J. Derr* and A. Rana, Virginia Tech, Virginia Beach (88).

ABSTRACT

Bulbs of various species are commonly planted in fall for their flower production in spring. Some of these species can be invasive, however, becoming a weed problem in landscape beds and turf areas. Field and container trials were conducted to determine control options for spring starflower [*Ipheion uniflorum* (Raf.) Traub.], Star-of-Bethlehem (*Ornithogalum umbellatum L.*), and grape hyacinth (*Muscari armeniacum* Leichtlin ex Baker). Sulfentrazone was applied either preemergence or postemergence at 0.25 and 0.38 lb ai/A and carfentrazone was applied postemergence at 0.015 or 0.03 lb ai/A. These treatments were compared to dicamba applied at 1.0 lb ae/A and to sulfentrazone at 0.25 lb ai/A plus dicamba applied at 1.0 lb ae/A. In container trials, PRE applications of sulfentrazone did not control these three species. A single POST application of sulfentrazone gave generally good control of Star-of-Bethlehem but unacceptable control of the other two species. A single application of carfentrazone provided fair to good control of Star-of-Bethlehem. Carfentrazone appeared to cause greater injury to spring starflower and grape hyacinth than sulfentrazone, although the control generally was still not acceptable. Dicamba applied alone provided generally poor control of the three species. Combining dicamba and sulfentrazone tend to improve control over either chemical applied alone. Sulfentrazone applied twice two weeks apart provided excellent control of Star-of-Bethlehem, poor to fair control of grape-hyacinth, and no control of spring starflower. Carfentrazone-ethyl applied twice two weeks apart provided good control of Star-of-Bethlehem, fair control of grape-hyacinth, and poor control of spring starflower. In field trials, sulfentrazone applied POST provide fair to excellent control of Star-of-Bethlehem, greater than that seen with carfentrazone-ethyl. Neither sulfentrazone nor carfentrazone-ethyl provided acceptable control of spring starflower and grape hyacinth. Sulfentrazone did not control any of the species following a PRE application. Control of invasive bulb species will depend on the specific species present, the herbicide chosen, and it application rate.

SEARCHING FOR PLANT ALLELOPATHIC COMPOUNDS IN THE SOIL MICROBIOME. S.M. Carver* and J. Kao-Kniffin, Cornell University, Ithaca, NY (89).

ABSTRACT

Microorganisms in the rhizosphere can attack or benefit the host plant over time through co-evolution to result in an enriched microbiome of organisms able to produce a vast array of natural products. Soil microorganisms provide an enormous genetic reservoir for biosynthetically produced allelopathic compounds. We present a method to isolate the alleopathic compounds produced by rhizosphere microorganisms, in this case, from the invasive plant, common ragweed, *Ambrosia artemisiifolia*. The method centers on metagenomic techniques common to molecular biologists to isolate environmental DNA, placing of the DNA into a vector-host system, and screening the clone library for positive or negative alleopathic activity. Screening was based on isolating clones that excrete small molecules. These clones were then extracted with methanol and the dried extract tested against two model plant species: *Poa annua* and *Lactuca sativa* in order to find products that could affect monocots or dicots. The effect of extracts was tested on seedling health phenotypes and several clones were found to have effects on the plants. Rhizosphere microorganisms could provide a novel frontier for isolating herbicidal or growth-enhancing compounds.

STRATEGIES FOR REDUCING PREEMERGENCE HERBICIDE INJURY TO BIGLEAF HYDRANGEA. J.C. Neal* and C.D. Harlow, North Carolina State University, Raleigh (90)

ABSTRACT

In response to increased popularity of bigleaf hydrangea (*Hydrangea macropylla)*, production of the species has grown in recent years. However, weed management remains an important challenge for producers because few herbicides are labeled for use on the species, and those which are labeled have been shown to cause crop injury. In 2008 a comparison of granular herbicides demonstrated significantly less crop injury from herbicides that did not contain a PPO inhibitor. In that experiment, injury from PPO inhibiting herbicides was significant but localized to tip necrosis. Within 6 weeks new growth had fully replaced the damaged tissues resulting in plants that were visually similar to non-treated plants. However, in subsequent research Freehand and Snapshot caused between 50% to 90% chlorosis or necrosis of the new buds. Because injury was localized to the growing points it is assumed that removal of granules from the growing points will prevent the primary cause of crop injury. Alternatively, based on the rates of plant recovery in following injury, it may be possible to “accept” the tip injury, prune the affected growing points and allow plants to re-grow. An experiment was established in 2014 to test three herbicide removal methods – irrigation after treatment, manual removal by shaking the plants, and pruning the affected buds after injury was visible. *Hydrangea macrophylla* ‘Endless Summer’ rooted cuttings were potted in 4-L pots using a pine bark substrate and top dressed with a standard slow-release nursery fertilizer. Plants were allowed to establish then were cut back to uniform height to stimulate branching. Treatments were applied when new growth was about 2 to 4 inches long. Granular herbicides included in the experiment were Freehand 1.75 GR @ 200 lb/A, Snapshot 2.5GR @ 200 lb/A, and Everris OH2 3G @ 100 lb/A. A spray treatment, Tower 6EC @ 32 fl oz/A, was also included for comparison. Injury was evaluated using several visual scales for tip necrosis, plant injury compared to non-treated controls, plant quality, and root growth. Additionally canopy height and width were measured 3 times and final above ground fresh weights were recorded. All granular treatments caused significant tip necrosis when not physically removed. Shaking plants to remove granules eliminated foliar injury from Freehand and Snapshot, but not OH2. Following pruning, plants with herbicide induced tip necrosis did not re-grow as rapidly as did non-treated plants. In this experiment plants treated with Tower EC and irrigated immediately after treatment were not visually different from the non-treated plants, but fresh weights and canopy volume were slightly reduced. Overall, results from this and other experiments underscore the importance of using non-PPO modes of action and physically removing herbicide granules from the growing points of *Hydrangea macrophylla* to avoid crop injury. Additionally, spray applications of Tower herbicide may provide an alternative if slight reduction in growth can be tolerated.

WEED CONTROL WITH BIOHERBICIDES IN NUSERY/LANDSCAPE BEDS. H.M. Mathers* and L.T. Case, Ohio State University, Columbus (91).

ABSTRACT

The research began on June 27, 2013 and evaluations were conducted at 37, 49, 65 and 105 DAT. A controlled release fertilizer (CRF) Osmocote Pro 17-5-11 fertilizer as a top-dress application was used in the field evaluations. Three by three foot plots /treatment /replications that were mulched with pine nuggets at 2" deep and three by three ft. of ½ cup of wood stove pellets plots/ treatment/ replication treated. Natural blow-in of weed seeds and the existing propagules of perennial weeds such as Quackgrass (*Elytrigia repens*), Canada thistle (*Cirsium arvense)* and bindweed (*Convolvulus arvensis*) occurred and exerted high weed pressures in all plots. Efficacy was rated on a scale of 0-10, with 10 being perfect weed control, 0 no control and $\geq$7 commercially acceptable. There were 13 treatments evaluated with five replications per treatment. Two commercial bio-herbicide products were used as comparisons, 20% acetic acid, at 10% v/v (Pharm Solutions Inc., Port Townsend, WA) and Iron HEDTA 26.52% (Neudorff North America) (W. Neudorff GmbH KG, Germany) at 10% v/v. Each of these and four alleopathic plant extracts produced at Ohio State University (OSU), Ohio State Innovation Foundation ("OSIF") designated as invention disclosure # 2013-036, Patent Pending # PCT/US2014/040344 were applied either directly onto the bark mulch already spread on the ground or the wood pellets (pre-treated at OSU). The objectives were to determine the efficacy and duration of weed control of different alleopathic plant extracts compared to other commercial organic herbicides treated on mulch or wood pellets. With ½ cup of wood pellets, only one treatment, the synthetic herbicide dimethenamid-p, 0.75% and pendimethalin, 1%) (BASF Corporation, Research Triangle Park, NC) at 37 DAT was providing commercially acceptable weed control. At 65 DAT even the synthethic with wood pellets was no longer commercially acceptable. This lack of control with the synthethic at 65 DAT speaks to the tremendous weed pressure in the Waterman Farm plots. All treatments with the 2" pine nugget at 37 and 49 DAT were commercially acceptable. At 65 DAT five treatments were commercially acceptable, four of which were OSU alleopathic plant extracts. The five treatments were Autumn olive 6% (*Eleagnus umbellata*), Tree of Heaven *6% (Ailanthus altissima*), Austrian pine 6% (*Pinus nigra*), Iron HEDTA 10% and Austrian Pine 5% + Black Walnut 5% (*Juglans nigra*). At 65 DAT all pellet treatments provided no control of weeds. By 105 DAT only three treatments evaluated provided efficacy ratings at or above commercially acceptable $\geq$7 OSU Autumn olive 6%, OSU Austrian pine 6% and OSU Austrian pine 5% + OSU Black walnut 5%.

UPDATE ON 2014 WEED SCIENCE RESEARCH IN THE IR-4 ORNAMENTAL HORTICULTURE PROGRAM. C. Palmer*, J. Baron, E. Vea, and E. Lurvey, IR-4 Project, Princeton, NJ (92).

ABSTRACT

In 2014, the IR-4 Ornamental Horticulture Research Program sponsored research on two weed science projects: 1) crop safety of over-the-top in-season herbicide applications, and 2) crop safety of over-the-top herbicides applied to ornamental grasses. The goal of the over-the-top herbicide application evaluations was to screen herbicides [Biathlon (oxyfluorfen + prodiamine), Dimension 2EW (Dithiopyr), F6875 4SC (sulfentrazone + prodiamine), Freehand G (dimethenamid-p + pendimethalin), Gallery SC (isoxaben), Marengo 0.0224%G (indaziflam), and Tower EC (dimethenamid-p)] for safety on woody and herbaceous perennials grown primarily in container nurseries with the emphasis on liquid herbicides.A pplications were made at dormancy and approximately 6 weeks later for all products. Biathlon was applied to 7 crops; Dimension EW was applied to 37 crops; F6875 4SC was applied to 13 crops; Freehand was tested on 3 crops; Gallery was applied to 2 crops; Marengo was tested on 4 crops and Tower was applied on 36 crops. The goal of the ornamental grass herbicide product evaluations was to determine crop injury with over-the-top applications. Three products were evaluated: Dimension 2EW (Dithiopyr), Gallery SC (isoxaben), and Pendulum G (pendimethalin). These products were applied to 18, 3, and 15 species, respectively. The results from this research will aid in the development of product labels and will help growers and landscape care professionals make more informed product choices.

SUBSTRATE TYPE AFFECTS CREEPING WOODSORREL GERMINATION IN NURSERY CONTAINERS. J. Altland*, USDA-ARS, Wooster, OH (93).

ABSTRACT

Container nursery substrates are composed primarily of softwood tree bark, with pine (*Pinus taeda*) bark being the predominant type used in the central and eastern United States. Bark is typically amended with various components including, but not limited to, sphagnum peatmoss, sand, compost, and other locally available agricultural or industrial byproducts. Sphagnum peatmoss is one of the most commonly used amendments, and is often incorporated at rates from 10% to 40% of the substrate volume. Sphagnum peatmoss can hold up to 20 times its weight in water, and thus is often used to increase the water holding capacity of pine bark substrates. Little is known about how sphagnum peatmoss, and other substrate components, affect weed establishment or herbicide efficacy in nursery containers. The objective of this research was to determine how pine bark substrate amended with sphagnum peatmoss affects creeping woodsorrel (*Oxalis corniculata*) germination in containers.

Black nursery containers (#3, approx. 12 L vol.) were filled with either 100% pine bark, 80 pine bark : 20 peatmoss, or 60 pine bark : 40 peatmoss (v:v). All substrates were amended with a commercial formulation of controlled release fertilizers to supply all macro and micronutrients. After potting, half of the containers were treated with 224 kg/ha Pendulum 2G (pendimethalin, BASF) and the other half were left untreated. A group of treated and untreated containers were seeded with 40 creeping woodsorrel seed the day following herbicide application (week 0) and a separate group of containers were seeded every 2 weeks thereafter for 8 weeks. Established creeping woodsorrel were counted 4 weeks after seeds were applied. Containers were placed within an overhead irrigation system and irrigated daily with approximately 0.6 cm of water. There were six single-pot replications per treatment and seeding date, arranged in a completely randomized design.

Increasing the ratio of sphagnum peatmoss in containers affected the substrate physical properties. Air space decreased and water holding capacity increased with increasing peatmoss level. Regardless of substrate type, Pendulum reduced creeping woodsorrel numbers in containers when seeds were applied up to 4 weeks after herbicide application. Among containers in which seed were applied 6 to 8 weeks after application, creeping woodsorrel numbers were similar in herbicide and non-treated containers. This suggests the herbicide barrier may have degraded to an ineffective concentration on the substrate surface in as little at 5 to 6 weeks. Substrate type had an inconsistent and relatively minor effect on creeping woodsorrel numbers. Among containers not treated with Pendulum, 100% bark had greater creeping woodsorrel numbers than the 60 : 40 substrate when seeds were applied 4, 6, or 8 weeks after potting. Among containers treated with Pendulum, weed counts were similar in all substrate types.

NEWSS YEAR-END FINANCIAL STATEMENT, 2013

NORTHEASTERN
WEED SCIENCE SOCIETY

WWW.NEWSS.ORG

AUDIT REPORT – Erin Hitchner

Financial documents and bank accounts were audited by Art Gover, Jim Steffel, and Todd Mervosh on January 6, 2014.

BEGINNING STATEMENT: As of October 31, 2012, the NEWSS checking account showed a balance of **$9,525.84** and the savings account showed a balance of **$31,765.71.** The 60 month CD balance was **$24,733.67** while the 18 month CD balance was **$10,090.80.**

ENDING STATEMENT: As of October 31, 2013, the NEWSS checking account showed a balance of **$7,710.19**, while the savings account balance was **$56,819.25**. The 60 month CD balance was **$25,274.37** and the 18 month CD balance was **$10,123.61**.

Net gain in assets for the fiscal year: $99,927.42 - $76,116.02 = $23,811.40.

FISCAL YEAR ACCOUNTING

New income as shown on bank statements for November 2012 through October 2013 including earned interest was **$55,798.70.**

New deductions from checking for November 2012 through October 2013 totaled **$31,987.30.**

*Net **gain** in assets: **$99,927.42 - $76,116.02 = $23,811.40.***

Non Annual Meeting Revenue/Expenses:

Non annual meeting revenue, which was comprised of interest earned as well as weed contest support from sustaining members ($5000.00) totaled $5597.70.

Non annual meeting expenses (Logo design fee, Dishonesty bond, website maintenance, weed contest travel, weed contest awards, administrative costs, WSSA Rep Travel, Contribution to Herbicide Resistance Planning Session, CAST dues and CAST Travel) totaled $17,633.55.
**Note – WSSA Dir. Science Policy Dues for both 2012 and 2013 were paid this fiscal year.*

Annual Meeting Revenue/Expenses:

Annual Meeting revenue (WSSA Joint Mtg payment, member dues, proceedings, BASF award, standing orders and sustaining dues) totaled $47301.00.

Annual meeting expenses (Proceedings, Treasurer Travel, Business Meeting, Abstract Fee, Standing Order Fee, Awards, Student Reimbursement and Photo Contest) totaled $14,353.75.

Education Fund Contributions

Contributions received for the Education Fund totaled $2,900.00. As of October 31, 2013, these funds were located in the NEWSS checking account.

NEWSS FISCAL STATEMENT

NOVEMBER 1, 2012- OCTOBER 31, 2013

Income Received	Amount	Comments
Annual Meeting - payment from WSSA	$ 35,000.00	Final
BASF awards	$ 300.00	Final
Interest earned on all accounts	$ 597.70	Final
Proceedings (Mail-in only - excludes ACTEVA purchases)	$ 360.00	Final
Membership renewal only	$ 300.00	Final
Standing Orders	$ 1,591.00	Final
Sustaining Membership: 17 to date	$ 9,750.00	Final
Sustaining Designated for Weed Contest	$ 5,000.00	Final
Education Fund Contributions (Currently in Checking Account)	$ 2,900.00	Final
Total NEWSS Income Nov 1, 2012 - October 31, 2013	***$ 55,798.70***	***Deposits + Interest***
Expenses Paid	**Amount**	**Comments**
Omni Press - Proceedings and Programs	$ 1,925.00	Final
Linda Rainville - Logo Design	$ 400.00	Final
CAST dues	$ 1,500.00	Final
Annual Meeting - Former Treasurer Expenses	$ 765.07	Final
Annual Meeting - NEWSS EC Lunch	$ 285.82	Final
October Business Meeting - NEWSS EC Dinner	$ 661.78	Final
EC Dishonesty Bond Insurance Renewal	$ 353.13	Final
Website maintenance	$ 200.00	Final
Webiste "A Small Orange" domain fee for NEWSS org.	$ 275.00	Final
Apex WebStudio LLC - WSSAAbstract setup fee	$ 300.00	Final
Standing Orders fee ($700 + postage)	$ 974.93	Final
Administrative (Postage, Proceedings mailing, etc)	$ 591.66	Final
WSSA Rep Trave to Summer Board Meeting	$ 750.00	Final
Annual meeting Membership Awards - Etch Art	$ 519.00	Final
Annual Meeting Student Reimbursement	$ 7,872.15	Final
Annual Meeting Student Awards (AM & PM Session)	$ 900.00	Final
Annual Meeting Photo Contest (1st and 2nd place only)	$ 150.00	Final
Weed NEWSS Contest 2013 Awards - Etch-Art	$ 678.12	Final
Collegiate Weed Contest host fee (funded travel 4 teams)	$ 1,320.00	Final
WSSA Dir. Science Policy YEAR 2012 Dues	$ 4,961.00	Final
WSSA Dir. Science Policy YEAR 2013 Dues	$ 4,961.00	Final
Contribution Herbicide Resistance Planning Session 2014	$ 1,000.00	Final
CAST Rep Travel	$ 643.64	Final
Total NEWSS Expenses paid Nov 1, 2012 - October 31, 2013	***$ 31,987.30***	***Final***

NORTHEASTERN
WEED SCIENCE SOCIETY
WWW.NEWSS.ORG

NEWSS FISCAL STATEMENT cont'd

NOVEMBER 1, 2012- OCTOBER 31, 2013

Total Fiscal Year Summary		
Total Income	$	55,798.70
Total Expenses	$	31,987.30
Net Gain	**$**	**23,811.40**

NEWSS Account Balances October 31, 2013	
Checking	$7,710.19
Savings	$56,819.25
60 Month CD (Meeting in reserve - Mar 2016)	$25,274.37
18 Month CD (Weed Contest - Mar 2014)	$10,123.61
Total Net Assets	**$99,927.42**

Signature of Preparer: [signature] Date: 1/6/2014

Accounts were reviewed by the undersigned and are in order.

Jim Steffel [signature] Date 01/06/14

Art Gover [signature] Date 1/6/14

Todd Mervosh [signature] Date 1/7/14

NORTHEASTERN WEED SCIENCE SOCIETY

NEWSS 2013 ANNUAL REPORT

Submitted by the Executive Committee
For the 68th Annual Meeting, January 8, 2014
Sheraton Society Hill, Philadelphia, PA

President's Report – Dwight Lingenfelter

The first ever joint meeting between WSSA and NEWSS was held on February 4-7, 2013 at the Hilton Baltimore. The 67th annual meeting of NEWSS and the 53rd annual meeting of WSSA was very successful. The two societies worked well together and most of the sessions were seamlessly combined yet we were able to maintain our identity during certain sessions. Thanks to all from our Executive Committee and to those from WSSA, namely Jim Kells, Joyce Lancaster, and Kate Counter, whose efforts made this a worthwhile event. The attendance at the overall meeting was 508, and of those, about 135 were affiliated with NEWSS. We again boasted a very successful NEWSS student contest in which 24 students from our society participated in the oral contest and 12 were involved in the poster session. Dispersed among the topical/commodity sections there were several symposia including: 1) Herbicide Application in the 21st Century; 2) Herbicide Resistance Management – The Path Forward; 3) Re-aligning your Course to Improve Student Learning; 4) Environmental Assessment of Herbicide Resistance; and 5) Holistic Invasive Weed Management: Integration of Science, Regulation, and Public/Private Policy. In addition to the meeting sessions, there were two social events including our popular nightcap/dessert gathering. New members to the Executive Committee were Rakesh Chandran as Vice President, Art Gover as Editor, Todd Mervosh as our Public Relations Representative, Bryan Dillahey as our Research & Education Coordinator, and John Orlowski as Grad Student Representative. Erin Hitcher, Treasurer and Keith Burnell, Membership Chair completed their first full year in these revamped positions. Work began soon after the meeting for the Executive Committee to plan another conference, schedule another joint Collegiate Weed Contest with NCWSS, and continue moving the society forward.

As stated in our Manual of Operating Procedures (MOPs), Executive Committee board meetings were scheduled, agendas circulated and hotel arrangements made. Committee Lists were updated, reviewed, approved and posted on the website. Proceedings are also posted on the web site one year after printing. Changes and modifications were submitted to the MOPs as needed.

An invitation was again offered to the Northeastern American Society of Horticultural Science to hold our meetings jointly in January 2014 and they accepted. In 2014, we are physically holding our poster session jointly and will be having a workshop that was developed by both societies. The NEWSS EC and two representatives from NE-ASHS met to review the hotel space in

October 2013 and final arrangements were made for the social, coffee breaks, and audio-visual set ups with the Sheraton Hotel staff. We were able to make our minimum number of room nights as of the hotel room cut-off date. As of this writing, we have exceeded our room night minimum. Renee Keese successfully solicited sustaining members for a contribution for the dessert and nightcap social, which will be held on Wednesday night. This open event continues to be very popular with the membership.

Our committee has finally completed its Education fund bylaws and they have been voted upon and approved by the NEWSS Executive committee. Funds are currently being collected.

As a result of last year's poll to select a new NEWSS logo, we are in the process of transitioning our emblem to the new design. It is a fresh looking logo that will be used on letterhead, websites, banners, plaques, apparel, and the like. We are in the process of getting a new banner to display at various meetings and events.

No Resolutions were brought forward by the Resolutions Committee in 2013. We have an excellent Vice-President candidate for 2014, Shawn Askew, who has previously served as WSSA Representative on our executive committee and has been very active as a Weed Contest Coach for many years. My sincere thanks to the entire NEWSS Executive Committee; you are a dedicated and hard-working group of volunteers dedicated to the realm of weed science.

President-Elect - Greg Armel

2014 meeting. Our meeting will be held at the Sheraton Society Hill on the week of Jan 6th 2014. We have a great program scheduled in conjunction with the Northeastern American Society for Horticultural Sciences (NEASHS). The theme of our 2014 meeting will be "Weed Control in the Future" which will be highlighted by a symposium with several special guest speakers from university, industry, and government who will provide us updates on the latest and greatest technologies in weed management and moreover give us some keys insights into the challenges in bringing new innovations to the market. Also, thanks to leadership from Vice President Rakesh Chandran we will be having a workshop focusing on strategies and insights for vegetable production. Finally, our regular program is also really exciting with well over 100 separate poster and oral presentations in agronomy, turfgrass and plant growth regulators, fruits and vegetables, ornamentals, weed ecology and biology, and vegetation management and restoration.

2015 meeting. I initiated some polling to determine our 2015 location for the NEWSS meeting. Through multiple rounds of surveying, we had identified two locations: Raleigh, NC and Orlando, FL. However, we could not identify a hotel in these two areas that would deliver us a room rate consistent with our past meetings and offer us the amenities we desire as an organization. To that end, I expanded my search to other locations and have found hotels in Roanoke, VA and Williamsburg, VA that have offered to host our group and have provided us rates consistent with our previous meetings. In any case, we will be making decisions on a hotel hopefully before our 2014 meeting.

Potential 2016-2018 joint meeting. I contacted Northeast Division of the American Phytopathological Society, Northeastern American Society of Agronomy, Eastern Branch of the Entomological Society of America, and the Northeastern America Society of Horticultural Sciences about doing a joint meeting that would be centered about Crop

Protection/Crop Health. The earliest this meeting could be done would be 2016, however, due to planning considerations and getting approval from our respective memberships it may take a few years longer to bring this together. The initial invite was warmly received by all of these societies and the main questions remain of where to host the event, what week in the year do we select, and how do we organize such a large gathering of diverse groups. I believe by doing an event like this, we raise the profile of our respective discipline and offer those interested in our ideas the opportunity to expand their knowledge and meet with diverse audiences for a better educational experience. In addition, there are always potential costs savings and other synergies that come forth from meeting in larger numbers. I look forward to working with the board in the coming years on this potential meeting.

Education fund (a.k.a the Endowment fund). Our committee has finally completed its Education fund bylaws and they have been voted upon and approved by the NEWSS Executive committee. Currently our Education fund has assets of $8,900 ($5,000 contributed by the board with $3,900 being submitted by generous NEWSS members). Our by-laws dictate that the fund must have at least $20,000 to become capable of delivering any type of disbursements for educational support. So right now we need to collect another $11,900 to activate our Education fund. The committee has challenged each member to donate at least $75 to contribute to our endowment fund so we can start to make a positive impact in the expansion of educational programs for the NEWSS. It is our hope that by the end of our 2014 meeting, we will have identified additional monies to activate this fund for society educational purposes.

Vice President - Rakesh Chandran

The assignments for the Society's committees were finalized, and are reported in the Proceedings.

Future Meetings: President-Elect Greg Armel has worked diligently to put together an excellent program for the 68th annual meeting in Philadelphia 2014, with assistance from the Northeastern American Society for Horticultural Sciences (NE-ASHS), and the section chairs of our society. A symposium on innovations in weed management and a joint-workshop focusing on weed management in vegetables are in place. I look forward to serving as the program chair for the 69th annual meeting in Roanoke, Virginia. Greg Armel has been in contact with sister societies in the region to consider the possibility of a joint-meeting that would cover all areas of crop protection (see President Elect's report). The outcome of discussions in the next few months will help determine the location and dates for our 2016 meeting.

Secretary/Treasurer – Erin Hitchner

Banking Information: Checking and savings accounts have been transitioned to Bank of America in order for Treasurer to have direct access to a local banking center. The 12 month Weed Contest CD ($10,123.61 as of October 31, 2013) still remains at the First Federal Fredericksburg Bank in PA until it matures in March of 2014. At that time, the additional $5,000 voted back in in 2012 to be added to this account will be added, and this CD will be transferred to the Bank of America account. The other CD, designated as our Annual Meeting in reserve, will mature in March of 2016. The balance for this CD is

$25,274.73 as of October 31, 2013. This CD will remain at the First Federal Fredericksburg Bank until that time to avoid penalty fees. After maturing, it too will be transferred to Bank of America.

The Treasurer, Erin Hitchner, serves as the primary signatory on the new accounts at Bank of America, while Keith Burnell, Membership Chair, serves as the secondary signatory.

Updated NEWSS Address:

Northeastern Weed Science Society
Erin Hitchner, Treasurer
P.O. Box 25
Woodstown, NJ 08098

Insurance: The Society maintains a Dishonesty Bond Insurance Policy with Fingar Insurance of New York. Due to a gradual increase in Society funds, this bond was raised from $50,000.00 to $100,000 for an additional fee of $96.13 (total annual premium of $353.13). Currently, the Society does not hold a Liability Policy, instead, when necessary, liability policies are purchased when they are need to single events (Weed Contest, short courses, etc).

Standing Orders: Irene T. Bradley continues to process and ship Standing Orders for Proceedings. In addition, past Proceedings are scanned and archived on the NEWSS website.

Website Maintenance: Rob Dickerson at Penn State University continues to maintain and update the NEWSS website. Executive Committee members work with Rob to post NEWSS updates related to contacts, annual meeting announcements and other society information.

Weed Contest: The Weed Contest was held this year in conjunction with the North Central Weed Science. To help offset the costs associated with traveling to this event, each team that participated received a travel allowance of $750. Four teams participated in the Contest, so a total of $3,000 were dispersed to the participating teams. Monies received by NEWSS for the Contest from Sustaining Platinum and Gold Members totaled $5,000, so $2,000 will carry over into 2014 for costs associated with the 2014 Weed Contest. The Weed Contest CD will mature in March of 2014 and will be used to cover Weed Contest expenses that exceed the annual contributions received from the Platinum and Gold level Sustaining contributions.

Annual Meeting: Registration fees and forms were approved by the EC and posted to the NEWSS website in November. Due to payment issues with ACTEVA, the decision was made to transition to PayPal for registration, membership renewal, Proceeding orders as well as Education Fund contributions. Also, for the first time, credit cards will be accepted at the Annual Meeting for registration, Proceedings orders as well as contributions to the Education Fund.

Hotel Tax Exempt Form: The NEWSS currently has a valid tax exempt certificate for PA (expires Feb 2015), MA (expires Oct 2018) as well as MD (expires Sept 2017). A request for Virginia tax exemption has been filed for the 2015 Annual Meeting to be held in Roanoke, VA.

Education Fund: The Education Fund Committee developed a set of by-laws which were reviewed and approved by the Executive Committee at the October Meeting. In addition, the Committee chose an Investment Group to manage the funds received for the NEWSS Education Fund.

Membership Chair – Keith D. Burnell

Membership update: As of January 2013, there are 324 members-at-large in the Updated Directory (2012-2013 Meeting Attendees) and additional contacts in the NEWSS Listserve who do not consider themselves regular members of the society. Members-at-large who have not been active in the society in the past 3 years are removed from the printed directory if not registered at the next annual meeting, or if they have not paid member dues only ($50.00) for the current fiscal year, or have since passed away.

NEWSS Annual Meeting	2005	2006	2007	2008	2009	2010	2011	2012	2013
Distinguished/Fellow	11	4	12	11	11	8	11	2	3
Dues only	Data not available				6	12	11	9	Data not available
Invited Speakers	Data not available	3	3	5	3	7	7	5	Data not available
Regular Members	200	139	148	119	132	91	104	110	94
Retired or Retired and Distinguished/Fellow	7	5	4	1	3	3	3	6	3
Student members	50	23	19	21	32	19	21	29	26
Sustaining Companies	Data not available	16	22	22	13	16	21	17	18
Symposiums and Workshops	172	22	39	14	76	0	18	20	Data not available
TOTAL	314	212	247	193	276	156	196	197**	144*
Proceedings Distributed	118	> 116	79	65	75	50	78	97	40
Standing Order Sales	50	50	50	50	50	50	55	46	43

*2013 Total = all in the 2013 column and number can be a double count of some member via Regular members possibly being part of the Sustaining Companies numbers. Also the numbers are from the WSSA/NEWSS combine meeting were many of the NEWSS member are/signed up as WSSA members.

Past-President – Antonio DiTommaso

The minutes from the 2013 Executive Committee meetings, annual business meeting, financial statements, newsletters, and other relevant publications were sent to the archives (Dan Kunkel) at the IR-4 Project Headquarters archive room in NJ.

The Awards Committee consisted of Antonio DiTommaso as chair, Mark VanGessel, Hilary Sandler, Dave Yarborough, and Jerry Baron. Nominations were received for Fellow, Award of Merit, Outstanding Researcher, Outstanding Educator, M. Garry Schnappinger Service Recognition, and Robert D. Sweet Outstanding Graduate Student (PhD only). Plaques were ordered for all award winners. The Awards program is being held during the opening session of the NEWSS meeting. NEWSS is publishing the Awards Brochure with pictures and biographies for all award winners.

The Past-Presidents' Breakfast is scheduled for Wednesday January 8, 2014 at the NEWSS annual meeting.

Editor - Art Gover

The Proceedings of the 68th Annual Meeting included 101 abstracts, out of 103 submitted titles. None of the submissions were in full-paper format, and one included a data table. We continued our relationship with OmniPress, who produced 150 copies, and shipped them to the meeting site.

We relied again on WSSA's submission site (wssaabstracts.com), managed by Apex Web Studios for our titles and abstracts. This site allows the Program Chair and Editor to develop program order, edit titles and abstracts, and order titles and develop author and keyword indices based on title-order. Now that I've had an opportunity to go through the process, the program and proceedings development process should be smoother for the 2015 and 2016 meetings.

I will seek to include an 'Instructions for Authors' link on the submission site, as it appears that as we progress to an entirely digital process with no actual need for the authors to layout their document, some elements of style are fading. Weed and crop nomenclature was the area of greatest non-compliance for the 2014 proceedings. It is my hope that more accessible Instructions, and the Editor reviewing submissions incrementally rather than as a batch in the future will result in more informative abstracts.

For 2014, I propose the Society consider transferring the ongoing archive of Society award winners to the NEWSS website and no longer including this in the proceedings. The Society Award and Collegiate Weed Science Contest winners from the year past are detailed in the proceedings. The NEWSS website would be a logical place to maintain the historic record of winners of each award, rather than devoting 22 pages of the Proceedings each year to an annual update.

Public Relations Chair – Todd Mervosh

The following is a summary of my activities as Public Relations Representative in 2013:

• In February at the combined Annual Meeting of WSSA and NEWSS in Baltimore, I took photos at the following events: General Session and Awards Program (photos of all speakers and award winners with award presenters), Opening Reception, Poster

Session, WSSA Business Meeting, NEWSS Past Presidents Breakfast (group pictures of Presidents), NEWSS Business Meeting (photos of President delivering speech, all award winners with presenters, passing of gavel from outgoing to incoming President, group photo of new Officers & Executive Committee), Dessert Social, and places & events of interest near the Hilton Baltimore (historic architecture, Camden Yards, thousands of Baltimore Ravens fans gathered for Super Bowl victory parade).

• Submitted an article about the Annual Meeting and photos of award winners to the editor of the WSSA Newsletter (article included in the April issue).

• In October at the NEWSS Board Meeting in Philadelphia, took photos of historic and other interesting sites near the Sheraton Society Hill for use in the November newsletter.

• Compiled, edited and distributed three issues of NEWSS NEWS (May, August, November) to the NEWSS listserve (over 300 email addresses) and for posting on the NEWSS website. Important components of each newsletter:

- May: Annual meeting recap and highlights; photos of all award winners; group photo and listing of NEWSS officers and executive committee members; committee chairs and members; plans for the Collegiate Weed Science Contest; summer weed tour dates; list of sustaining members for the Society.

- August: Summary of the weed contest including names of winning individuals and teams, plus a few general photos; call for NEWSS awards nominations plus criteria for each award; early information about the 2014 annual meeting in Philadelphia; instructions for submitting titles and abstracts.

- November: Lots of information about the upcoming annual meeting, including instructions for meeting registration and room reservations at the Sheraton Society Hill; hotel website, travel and tourist information for Philadelphia; condensed program schedule for the annual meeting; photo contest rules; candidate(s) for NEWSS Vice President; photos of weed contest award winners.

• Updated the NEWSS email listserve periodically, deleting invalid email addresses and adding new email addresses as needed.

• Emailed reminders about title and abstract submission deadlines in addition to annual meeting registration and hotel room reservation deadlines.

• Forwarded several emails to the listserve regarding job opportunities and other important announcements related to weed science.

Research and Education Coordinator – Bryan Dillehay

The Northeastern Weed Science Society (NEWSS) has regularly offered pesticide recertification and Certified Crop Adviser (CCA) credits for several years at the Annual Meeting. Last year, at the joint NEWSS and Weed Science Society of America (WSSA)annual meeting in Baltimore, credits were offered for North Carolina and all northeastern states except New York and New Hampshire. New York and New Hampshire credits were not offered due to the complexity of the application process for such a large meeting with numerous speakers. For the 12 states that pesticide recertification credits were offered, 43 recertification credits were awarded across the 14 sessions that were available. The average recertification credits awarded per state was 3.58 and the average recertification credits awarded per session was 3.1. Pennsylvania and North Carolina had the most recertification credits requested at 9 each while the most recertification credits were awarded in the Agronomic Crops session. There were no credits requested from Rhode Island, Virginia, Vermont, or West Virginia. 15 CCA credits were awarded at the meeting with an average of 1.1 per session and the most CCA credits were awarded in the Herbicide Application Symposium with 3.

Based on the trend of declining pesticide recertification requests of certain states and the increasing level of complexity in the application process, I have narrowed down the list of states where credits will be offered at the annual meeting. For the 2014 annual meeting, recertification credits will be offered for Connecticut, Delaware, Maryland, North Carolina, New Jersey, Pennsylvania, and West Virginia. CCA credits will be offered as well. Recertification and CCA credits will be offered in 12 sessions of the annual meeting.

At the October NEWSS Executive Committee (EC) meeting, the continued offering of recertification and CCA credits in light of the declining requests and complex application process was discussed. The EC decided it was best to continue to offer credits for those states where requests have been adequate and steady for the last 3 years, but to discontinue offering credits for states with very few requests over the last 3 years. The offering of recertification and CCA credits will continue to be evaluated for upcoming annual meetings.

Graduate Student Representative – John Orlowski

In July 2013, NEWSS graduate students traveled to Monmouth, Illinois for the joint NEWSS-NCWSS Collegiate Weed Contest. Participants from four NEWSS teams participated in the event and represented NEWSS admirably and had a good time as well. A professional-student matching event was proposed for the NEWSS meetings this spring, but little interest was expressed in the event. Moderators for the graduate student paper contest at the 2014 meetings will be John Orlowski of the University of Kentucky and Kate Venner from Virginia Tech. The next big event for the graduate students is the mixer at the 2014 meetings. Ice-breakers and speakers are still being planned and considered.

CAST Representative - Randy Prostak

<u>John Bonner Retires from CAST.</u> John Bonner looks to the future by pulling from the past. "For forty years CAST has been inspired by visionary thinkers such as Charles Black, Norman Borlaug, and Gale Buchanan," he said. "CAST's mission came as a charge from the National Academy of Science, and I feel confident that the organization, led by incoming EVP Linda Chimenti, will continue to inform the public and policymakers."

Those who have worked with Dr. Bonner know this is a typical tone from the man who has been CAST's EVP/CEO for the past seven and a half years. Even as he shifts into a life that will offer more time for family, hobbies, and whatever else he decides to take on, Bonner is still focused on how best to communicate science-based information about agriculture. In the photo at right, Bonner joins the respected ag communicator Orion Samuelson (left) to present an award to an FFA essay contest winner.

His departure has been carefully planned to enable a smooth transition. He announced the move several months ago, and this has given him time to work with Ms. Chimenti so that the change would be effective and efficient. Bonner considers communication, quality publications, and student programs as his key areas of focus, and he feels confident these will continue to be strengths at CAST.

Linda Chimenti New Executive Vice-President. The Board of Directors of the Council for Agricultural Science and Technology has appointed Linda Chimenti, current Chief Operating Officer (COO) of CAST, as the next Executive Vice President of the 40-year-old organization. Dr. John Bonner, current Executive Vice President/Chief Executive Officer (EVP/CEO), will leave that position at the end of January 2013 to pursue other interests.

Ms. Chimenti has years of experience as a CAST administrator, and she is committed to the organization's mission. As she says, "It is an honor to have the opportunity to take on a new role in CAST, an organization that I have served for twelve years with growing responsibilities." Ms. Chimenti also considers the move a compliment to the entire CAST staff. "The Board of Directors has expressed their confidence in our ability to maintain CAST's output of high quality, credible publications and programs."

After working at Iowa State University's Center for Agricultural and Rural Development, Ms. Chimenti joined the CAST organization in January 2001 as Managing Scientific Editor and has led the editorial department continuously since that time. During this period, she has been responsible for the production of approximately 140 CAST publications, including Task Force Reports, Issue Papers, CAST Commentaries, Annual Reports, and organizational histories.

In 2008, Ms. Chimenti was promoted to Director of Council Operations and added staff supervision and management duties to her responsibilities. She assisted the Board of Directors and Board of Representatives with a major restructuring of the organization's governance in 2009. Since 2010, Ms. Chimenti has served as COO, working closely with Dr. Bonner on initiatives to expand CAST's impact and broaden its audience.

"John Bonner has been an excellent mentor during the years we've worked together," says Ms. Chimenti. "He has provided team leadership for the staff and brought many new ideas for moving the CAST organization forward. He set a high standard for the next EVP to follow."

Since 2005, Dr. Bonner has worked extensively with the CAST Board of Directors and Board of Representatives to revitalize the organization, and to establish a Board of Trustees. Working closely with Ms. Chimenti, he implemented many new projects, including the production of videos based on CAST publications; the active use of social media to communicate with a wider audience; and a CAST Education Program to make materials available to university students and staff at land-grant institutions.

Dr. Bonner hosted rollout presentations of CAST publications at venues in Washington, D.C., and visited numerous congressional offices to familiarize staffers and legislators with CAST resources. Dr. Bonner also developed a major campaign emphasizing the "Three I's" concept: Input, Impact, and Investment. As he stated, "I am confident that

Linda is focused on the 'Three I's' concept, and she will do an excellent job as EVP of CAST."

According to Dr. Phil Stahlman, CAST President, "We are very pleased that this transition of leadership will take place within the CAST organization now that Dr. Bonner is stepping down to pursue other interests. We are deeply grateful for all he has done for CAST, and we wish him all the best. We welcome Ms. Chimenti to her new role and look forward to working with her in this capacity."

Melissa Sly becomes the Director of Council Operations.

New publication format: Ag quickCAST - Each one-page Ag quickCAST contains excerpted material from its companion CAST document in an easy-to-read format consisting of topic sentences and bulleted lists. Each document identifies experts to contact for more information and provides a link to the full publication. To introduce the new series, CAST has created Ag quickCASTs for three previously published documents; in the future, an Ag quickCAST will accompany each new CAST publication release.

David Songstad New CAST President-Elect. Dr. David Songstad was chosen as the President-Elect for 2013-2014. He will officially assume leadership responsibilities at the conclusion of CAST's Fall Board Meeting at the end of October. In 2014-2015, he will become the 42nd President of CAST, a singular honor and responsibility dating back to 1972, when Charles A. Black and Norman Borlaug, along with other committed scientists, spearheaded the movement to "bring science-based information to policymaking and the public."

2013 CAST publications

February 2013 - Spanish version of the February 2012 Issue Paper IP49: Herbicide-resistant Weeds Threaten Soil Conservation Gains: Finding a Balance for Soil and Farm Sustainability

Herbicides were developed during the twentieth century to be used with conventional tillage for weed control. Conservation (or minimum) tillage subsequently evolved, which enabled less soil damage when used with herbicides. Selection pressure, however, has resulted in weed species that have made adaptations for survival in conjunction with tillage. The U.S. government has put several federal policies and programs in place that help determine the selection and implementation of crops and conservation programs in relation to herbicides and tillage. This Issue Paper examines the impact of certain weed management practices on soil conservation objectives and addresses ways to mitigate negative effects. Chair: David R. Shaw, Office of Research and Economic Development, Mississippi State University.

March 2013 - Food, Fuel, and Plant Nutrient Use in the Future, Issue Papers - IP51

Future food, fiber, and fuel demands will not be met by expanding cropland area; continued advances in nutrient use efficiency will moderate increased nutrient demand. With a growing population, dwindling arable land, and an increased demand for biofuels, the world cannot count on an expansion of harvested area to fill the demands of the world's growing population. The use of genetics to improve crop productivity, promote soil conservation and management, and use nutrients efficiently is necessary. The key lies in supporting research and development in these areas. This CAST Issue

Paper looks at the background leading to the current situation and addresses the resulting requirements as world food production develops during the next 40 years. The authors use data to analyze factors influencing crop production now and indications of what is to come, and they show how research regarding nutrient use, recovery, and recycling is crucial. Chair: David Zilberman, University of California -- Berkeley.

June 2013 - Impact of the Precautionary Principle on Feeding Current and Future Generations, Issue Papers - IP52

This paper looks at the history and purpose of the precautionary principle (PP) and examines problems of ambiguity, arbitrary application, and bias against new technologies. Because the publication is especially focused on the need to feed a growing population, the case studies center on agricultural issues such as pesticide use, genetically modified foods, and food irradiation. The authors state that the PP has played an important part in bringing attention to appropriate risk management. If it is applied in its more stringent formulations, however, the PP will suppress innovation, to the detriment of both the economy and human health. Chair: Gary Marchant, Arizona State University.

September 2013 - Animal Feed vs. Human Food: Challenges and Opportunities in Sustaining Animal Agriculture Toward 2050, Issue Papers - IP53

A popular belief exists that animal agriculture competes for human food supplies and wastes resources. Little scientific information exists regarding this concept, and it is therefore necessary to quantify human-inedible feedstuff use and re-examine the feed efficiency concept using a ratio of human-edible feedstuff input to human-edible (animal protein) output. This paper addresses the commonly heard argument that livestock compete with humans for food resources, thus providing support for the continued existence of livestock production. It also addresses the knowledge gap that currently exists as to the quantity of human feed and fiber by-products used within animal agriculture. Chair: Jude Capper, Montana State University.

WSSA Representative - Jacob N. Barney

WSSA Executive Committee:

President	Jim Kells
President-Elect	Joe DiTomaso
Vice-President	Dallas Peterson
Past-President	Rod Lym
Secretary	John Madsen
Treasurer	Ian Burke
Executive Secretary	Joyce Lancaster
Director of Science Policy	Lee Van Wychen
Members at Large	Andrew Kniss, Les Glasgow, Sarah Ward, Dan Kunkel

Website: www.wssa.net

A list of items from 2013 of relevance and importance to NEWSS:

WSSA is looking at the creation of a USDA-NIFA Subject Matter Expert that is similar to the EPA SME that was previously held by Jill Schroeder. This person would be embedded in NIFA and work directly with Michael Bowers and other personnel

Mike Barrett is the new EPA Liaison

The new website is up and running, and content is managed by a new web content manager

The Herbicide Handbook is going digital. A digital database is being built to allow a web version first, and potentially a smartphone app

IPSM has it's first impact factor 0.98

All journal impact factors are steady or tracking up

We should encourage NEWSS members to participate in WSSA BOD activities and committees.

Upcoming Meetings:

2014 Vancouver, British Columbia, Canada

2015 Lexington, KY

2016 Puerto Rico

NORTHEASTERN
WEED SCIENCE SOCIETY

2014 NEWSS BUSINESS MEETING

Call to order – President

President Dwight Lingenfelter called the business meeting to order at 5:20 pm. He thanked the members and guests for attending the meeting this year.

Acceptance of the Minutes from the 2013 Annual Business Meeting

Copies of minutes were made available and also projected in the meeting room. Todd Mervosh pointed out a correction (John Lydon in Necrology report was incorrectly spelt). Brian Olson moved to accept the minutes and Rich Bonnano seconded the minutes as amended. The motion passed unanimously.

Necrology Report - Membership Chair

Keith Burnell reported the loss of several NEWSS members during the 2013 year which included Bill O'Neil (November 12th), John Arthur Meade (April 17th), A.J. Powell (). Keith also provided brief obituaries and the survivorships of the members who passed away.

Executive Committee Reports

President D. Lingenfelter indicated that the executive committee members have submitted their reports which have been compiled and place for review at the registration desk.

Herbicide Resistance Summit Report

Mark VanGessel had represented the NEWSS at the Herbicide Resistance Summit for which the society had contributed $1000.00. Mark was requested by the president to provide its membership with an overview of the summit. It was second of the series of such summits held in 2013 at the National Academy of Sciences in September to draw the attention of a wider audience and take a look closer into the social barriers in resistance management. Closer work with rural sociologists, economists was need to understand this and put the science to practice. A subsequent larger summit/symposium to follow up was expected to be planned at the WSSA meetings this year (this meeting was attended by about 100 people).

Research and Education Committee Update

Bryan Dillehay, Chair of Research and Education committee expressed concerns about the declining signups for Recertification credits at the meeting and about certain states not having any signups at all during the past several years. Based on the efforts of the society to offer this service, it is questionable whether this should be continued. Bryan had therefore applied for credits only from 7 states in 2014 meetings (instead of 14 previously) and will be monitoring the numbers closely to see if needs to be continued.

Bryan felt that the our efforts may be better used in a different area based on the time commitments required for this process and marginal returns. Bryan solicited feedback from the society about the proposed changes.

CAST Representative Update

Randy Prostak provided an overview about the membership of CAST and its mission to assemble, interpret and foster a multi-disciplinary approach to deliver science based information in Agriculture. NEWSS was a member society since its inception in 1972. There are 16 societies including WSSA, NCWSS, WWSS, and APMS, and 7 university members. The abundance of "misinformation" in the digital age are various activists are perceived to be a challenge for CAST. A proposal by Randy to disseminate educational resources to individual members of member societies is being considered. Various avenues that CAST use to deliver information were listed. A 1-sheet summary of 20-page papers will be made available based on request by the plant group mainly to be passed on to legislature. The groups are Animal Science, Food Science, Plant Science (that NEWSS is on). Randy pointed out that Tri-Societies, APS and ESA do not have memberships. Based on Randy's suggestion CAST decided to form a small committee in the NEWSS to solicit ideas and issues expected to become concerns in Agriculture in the region so that CAST can be more proactive. The members of this committee would be Jeff Derr, Brian Olsen, Dave Mayonado, Lee Van Wychen, Bill Curran, Rich Bonnano, and Matt Ryan. Randy encouraged the society members to seek information from CAST and Randy promised to help provide any necessary information.

Treasurer's Update

Treasurer Erin Hitchner provided an updated about the society's financial status. Erin noted that the society in sound financial grounds. For the joint weed contest, $3000 was spent to offset student travel costs. Payments processed this year through Pay-Pal was considered to be more efficient and cost-effective for the society, compared to that my Acteva. Some issues with multiple registrations were encountered this year which is hoped to be resolved for next year. Erin filed for tax-exempt status from VA in preparation for next year's meeting. Income for the society from the past fiscal year was $55,798 and expenses were at $31,987, resulting a net gain of $23,811 during the fiscal year. The dues paid by WSSA ($35,000) as a result of the joint-meeting and reduction in expenses (by $12,000). Erin checked with the attorney hired to help with the Education Fund to rule out any potential problems associated with such net gains being a non-profit organization. No issues were expected and the funds could be used for any legitimate purposes related to added meeting expenses etc. Erin also provided an update on the Education Fund. The Education committee established bylaws which were approved by the EC. $5000 was contributed by the society to initiate this fund. Since then additional $4000 was contributed by various NEWSS members. Generous contributors included Rich Bonanno, Robin Bellinder, James Parochetti, and Scott R. Glenn. Donations may be given year round through Pay-Pal or sending a check to the Treasurer.

Audit Committee Report

Jim Steffel representing the audit committee reported that he went over the accounts with Erin Hitchner and confirmed that all the numbers matched and are in good order were approved and signed by other members of the committee as well. He lauded the

Treasurer for her efforts. Prasanta Bhowmik move to approve the audit committee's report and Mark VanGessel seconding the motion. The motion passed unanimously.

Archives Committee Report

Dan Kunkel reported that he had received the needed updates for the archives from Toni DiTommaso (?) and that the will go into the archives next week.

Weed Science Policy Update

Director of Science Lee Van Wychen thanked the society for having provided guidance on a number of issues related to Weed Science. He reported that the USDA-ARS position for Program Leader in Weed Science may be a non- Weed Scientist. He urged the membership to contact potential decision makers about this concern. Lee also brought up the concerns regarding the passing of the Farm Bill. He mentioned that the discretionary funds associated with the budget was reinstated. Lee provided an update for the RNAi technology and the regulatory framework with which EPA will approach it. A committee is expected to take a closer look at this during a Jan 28th meeting and is expected to get faster approval as a bio-pesticide through FIFRA, based on the safety and soundness of the science behind the technology. The proposed labelling of GM products is being opposed by the American Phytopathological Society. The Science Policy Committee with WSSA also recommended a similar approach and he expected the regional societies to pass resolutions based on this. The final approval will made at the Vancouver meeting by WSSA Board. Any comments from the membership regarding this may be passed on to Lee.

Awards Committee – Past-President

President Dwight Lingenfelter announced the winners of the Joint Collegiate Weed Contest held in Illinois. The NE region had four Universities that participated (NC State, Cornell, Guelph, and Penn State). The following are the winners:

Graduate Student Division:

First place – NC State Team A:
Thierry Besancon, Sushila Chaudhari,
Alexandra Knight, R. Eric Paynter

Second place – NC State Team B:
Nicholas Basinger, Lewis Braswell,
Bryan Hicks, Matthew Inman

Third Place – Penn State University
Kelly Patches, Denyse Schrenker, Clair Keene

Individual Awards (Graduate Student Division):

1st Place – Suchila Chaudhari – NC State University
2nd Place – Courtney Stokes – Cornell University
3rd Place – Clair Keene – Penn State University

Under-Graduate Student Division:
First Place - Univ. of Guelph Team A:

Jacob Nederend, Jordan Eyamie, Johanna Burrows

Second Place - Univ. of Guelph Team B:
Thomas Heeman, Nicole Bloetjes, Joshua Burrows

Third Place – Penn State University
Lyle Stephens, Kaycee Browell,
Phil Sebulsky, Rachel DeFlumeri

Individual Awards (Under-Graduate Student Division):

1st Place – Johanna Burrows, Univ. Guelph
2nd Place – Nicole Bloetjes, Univ. Guelph
3rd Place – Jordan Eyamie, Univ. Guelpp

Graduate Student Presentations. Past-President Toni DiTommaso presented the winners of the Graduate Student Presentation in the 2014 meeting. Toni thanked the judges who served on the committee and BASF for presenting the monetary awards for the winners. Toni mentioned that overall the presentations were of excellent quality and that students maintained good eye-contact but urged students to refer to the guidelines while preparing the presentations. Toni indicated that there were 22 students who competed in this year's contest (in some instances the hypothesis was not addressed, objectives not stated, photos, statistics unclear, unexplained acronyms). Comments in evaluation sheets were brief due to time constraints. Scores were really close among winners. The winners were:

2nd place: C. L. Keene, W.S. Curran, M. Dempsey, Penn State University, for the paper titled "Corn, soybean, and wheat performance in an organic rotational no-till system in Pennsylvania".
1st Place: M. Elmore, J.T. Brosman, G.K. Breeden, J.J. Vargas, and G.R. Armel, Univ. Tennessee, for the paper titled "Combinations of topramezone and cloquintocet-mexyl for creeping bentgrass (*Agrostis stolonifera*) and grassy weeds".
Honorable Mention: K.M. Patches and W.S. Curran, Penn State University.

Research Poster contest – Poster Awards Committee Chair Keith Burnell presented the awards for student research posters. He thanked the judges T. Mervosh and D. Ganske for their help with judging the posters. Students were urged to present their own work, speak up to the judges, and present the information in simple terms, avoid too much information in the posters, show/follow up on statistics, use correct poster size, appropriate bullets, ensure quality of graphics, literature cited, etc (follow criteria). The winners are:

2nd place (tie): Z. R. Taylor, W.J. Everman, NC State University, "Novel herbicide combinations for HPPD inhibitors with plant growth regulators for improving parlmer amaranth control".
2nd place (tie): M.C. Cox, E. Kollakova, S.D. Askew, Virginia Tech, "Mobility and metabolism of oxadiazon in two biotypes of goosegrass"

1st place: B. A. Hicks, W.J. Everman, NC State University, "Tolerance of *Arundo donax* grown as a bioenergy crop to applications of postemergence herbicides.

Photo contest – Erin Hitchner committee chair presented the awards (from 24 submisions). The winners are:

3rd Place: J. Cote ?
2nd Place: W. S. Curran – Bull thistle (*Cirsium vulgare*) inflorescence
1st Place: A. Gover – Japanese beetles on invasive knotweed (*Polygonum)* species

Old Business - None

Officer Changeover and Presentation of the Gavel

Dwight Lingenfelter reminisced his student days and encouraged students to step up to the plate whenever necessary to ensure the smooth functioning of this friendly society and eventually their work will be recognized. He then transferred the office of President over the Greg Armel who in turn provided Dwight a plaque commemorating his great service to the NEWSS.

Resolutions - committee chair

President Greg Armel, moving on to new business, asked for any new resolutions. Andy Senesac chair of resolutions committee stated that there were no new resolutions. The committee chair, Andy along with members Courtney Stokes and Jennifer D'Appollonio-Cote was requested to serve for another year and they agreed to do so. Greg appreciated their efforts.

Nominations committee

Greg Armel requested Mark VanGessel chair of Nominations committee to present nominations. Mark thanked the nominating committee which comprised of Larissa Smith, Bill Curran, Sudeep Mathews for their assistance. The committee communicated electronically to come up with an appropriate candidate for Vice President and they put forward Shawn Askew for this position. Greg Armel asked the membership for additional nominations. There were no additional nominations from the floor. Motion was made by Mark VanGessel to close and was seconded by Toni DiTommaso. There was no further discussion and Shawn Askew was elected Vice President unanimously.

Greg nominated Toni DiTommaso to serve as the chair of the Nominating committee this year. He appointed Javier Vargas at large and requested three other members to be nominated from the floor. Rakesh Chandran nominated Rob Richardson, Randy Prostak nominated ? Johnson (?), and Dan Kunkel nominated Todd Davis. The nominations were unanimously approved by the membership present.

Weed Contest Site and Date

Dwight Lingenfelter stated that the next collegiate weed contest will be hosted by Penn State by Biil Curran, D. Lingenfelter, and the farm manager, to be held in late July.

Next NEWSS Meeting Site

Greg Armel stated that the site for 2015 meetings is not firm yet. Negotiations with Hotel Roanoke does not appear to be working out and he is currently discussing with Williamsburg Lodge which looks promising. A final decision has not been made yet. Greg also urged the members to consider making contributions to the Endowment Funds. He thanked Renee Keese and Jeff Derr for initiating this effort and also Erin Hitchner for taking care of all the paperwork and details to ensure a successful launch.

Presentation of the Executive Committee

President Greg Armel presented the new NEWSS Executive committee to the membership including the new WSSA representative Prasanta Bhowmik.

Adjourn

Mark VanGessel moved to close the business meeting and Dan Kunkel seconded the motion. The motion carried unanimously and the meeting ended at 6:38 pm.

NEWSS MEMBERSHIP DIRECTORY

Last Name	First Name	Organization	Phone No.	Email Address
Abbey	Tim	Penn State Cooperative Extension	717-840-7408	tma13@psu.edu
Ackley	Bruce	The Ohio State University	740-225-2014	ackley.19@osu.edu
Agnew	Michael	Syngenta Crop Protection	610-444-2063	michael.agnew@syngenta.com
Alea	Stephanie	Rutgers University	732-932-9711 x116	stephya@eden.rutgers.edu
Allen	Jayla	Bayer Crop Science		jayla.allen@bayer.com
Armel	Greg	BASF Corp.	919-547-2674	gregory.armel@basf.com
Arsenovic	Marija	Rutgers - IR-4 Headquarters	732-932-9575 x 4609	arsenovic@aesop.rutgers.edu
Ashley	James	Evonik Goldschmidt Corporation	804-452-5692	james.ashley@evonik.com
Askew	Shawn	Virginia Tech	540-231-5807	saskew@vt.edu
Atland	James	USDA-ARS ATRU	330-263-3870	James.altland@ars.usda.gov
Averill	Kristine	Penn State University	860-248-9969	kma236@psu.edu
Baker	Robert	Scotts Company	937-645-2628	robert.baker@scotts.com
Barcel	David	OHP Inc.	262-392-3004	dbarcel@ohp.com
Barkman	Gary	Montgomery Weed Control Inc.	301-503-6024	gbarkmansr@gmail.com
Barney	Jacob	Virginia Tech	540-231-6323	jnbarney@vt.edu
Barolli	Sali	Imperial Nurseries	860-653-1509	saizba@yahoo.com
Baron	Jerry	Rutgers - IR-4 Headquarters	732-932-9575 x4605	jbaron@aesop.rutgers.edu
Barrett	Mike	Argyle Country Club	301-774-5102	turfgod1@CS.com
Bates	Ryan	Penn State University	814-863-1014	rtb175@psu.edu
Batts	Roger	North Carolina State University - IR-4 Center	919-515-1668	roger_batts@ncsu.edu
Baxter	David	DuPont Crop Protection	302-366-5065	david.a.baxter@usa.dupont.com
Becker	Chris	BAAR Scientific LLC	607-342-3610	becker89@fltg.net
Bellinder	Robin	Cornell University	607-255-7890	rrb3@cornell.edu
Beste	Ed	Univ. Maryland, Salisbury Facility (Emeritus)	410-742-8788	ebeste@umd.edu
Bhowmik	Prasanta	University of Massachusetts	413-545-5223	pbhowmik@pssci.umass.edu
Bonanno	Richard	University of Massachusetts	978-361-5650	rbonanno@umext.umass.edu
Borger	Jeffrey	Penn State University	814-865-3005	jab267@psu.edu
Bostic	Heidi	University of Tennessee	865-974-7324	hbostic@utk.edu
Bowe	Steve	BASF Corp.	919-547-2559	steven.bowe@basf.com
Bowers	Ian	Pennsylvania Dept Agriculture	814-323-7182	bowers.ian29@gmail.com
Bradley	Irene	NEWSS Standing Orders	631-727-3595	it21@cornell.edu
Bravo	Melissa	Meadow Lake Farm Consulting Services	814-574-4067	bravomelissa@yahoo.com
Breeden	Gregory	University of Tennessee	865-974-7208	gbreeden@utk.edu
Brill	Nancy	Syngenta	919-306-6094	Nancy.Brill@syngenta.com
Brooks	Thomas	Crop Management Strategies	610-767-1944	tjb_cms@fast.net
Brosnan	James	University of Tennessee	865-974-8603	jbrosnan@utk.edu
Buck	Jenny	Wildflower Photographer	732-892-5507	jebuck_1@yahoo.com

Buck	Elizabeth	Cornell University	607- 425-3494	emb273@cornell.edu
Burch	Patrick	Dow AgroSciences LLC	540-382-3062	plburch@dow.com
Burnell	Keith	Syngenta Crop Protection	315-209-7580	keith.burnell@syngenta.com
Cain	Nancy	Cain Vegetation Inc.	519-853-3081	cain.vegetation@gmail.com
Calabro	Jill	Valent USA Corporation	443-235-0612	jill.calabro@valent.com
Cardina	John	Ohio State University	330-263-3644	cardina.2@osu.edu
Case	Luke	The Ohio State University	614-292-0209	case.49@osu.edu
Chamberlin	Joe	Valent USA Corporation	770-985-0303	jcham@valent.com
Chandran	Rakesh	West Virginia University	304-293-6131	rschandran@mail.wvu.edu
Chism	Bill	US EPA, OPP	703-308-8136	chism.bill@epa.gov
Coffman	Benjamin		443-547-1225	bencoffman41@yahoo.com
Corbett	Jerry	Quali-Pro	919-279-2584	jerryc@quali-pro.com
Cote	Jennifer	University of Maine	207-581-2924	jennifer.dappollonio@maine.edu
Cox	Michael	Virginia Tech	540-231-5835	mcc83@vt.edu
Cranmer	John	Valent USA Corporation	919-387-2013	jcran@valent.com
Crockett	Benjamin	Penn State University	814-863-7112	bcc158@psu.edu
Cummins	John	University of Tennessee	865-228-9789	recon77@utk.edu
Curran	William	Penn State University	814-863-1014	wcurran@psu.edu
Cushman	Mark	The Scott Company	937-644-0011	mark.cushman@scotts.com
Custis	Gary	PBI Gordon Corporation	816-460-6215	gcustis@pbigordon.com
Cutulle	Matthew	Syngenta	772-794-7170.	Matt.Cutulle@syngenta.com
David	Paul	Gowan Company	717-560-8352	pdavid@gowanco.com
Davis	James	Vegetation Managers Inc.	814-765-5875	vminc1@verizon.net
Davis	Todd	Delaware Dept. Agriculture	302-697-4468	Todd.Davis@state.de.us
Davis	Vince	University of Illinois	217-333-1531	davisv@illinois.edu
Derhoeden	Peter	University of Maryland	301-405-1337	pd@umd.edu
Derr	Jeffrey	Virginia Tech	757-363-3912	jderr@vt.edu
Dillehay	Bryan	Monsanto Company	814-404-2683	bryan.dillehay@monsanto.com
Dillon	Cory	Penn State University	814-863-1014	csd109@psu.edu
DiTommaso	Antonio	Cornell University	607-254-4702	ad97@cornell.edu
Dobbs	Jeffrey	OHP Inc.	770-992-0121	jdobbs@ohp.com
Dougherty	Ryan	Virginia Tech	540-231-6323	dougherty@vt.edu
Druist	Dervin	Syngenta	717-360-8322	dervin.druist@syngenta.com
Dufoe	Allan	FMC Corporation	410-570-6206	allan.d.dufoe@fmc.com
Dunst	Richard	Cornell Lake Erie Research and Extension Lab	716-785-9910	rdunst@netsync.net
Dutt	Timothy	LABServices	610-562-5055	tedutt@ptd.net
Egan	Franklin	Penn State University	814-865-6679	jfe121@psu.edu
Ekins	Rick	FMC Professional Solutions	215-299-5836	rick.ekins@fmc.com
Elmore	Matthew	University of Tennessee	865-974-7324	melmore6@utk.edu
Estes	Tony	United Phosphorus Inc.	864-202-7526	tony.estes@uniphos.com
Evans	Glenn	Cornell University	607-255-9085	gje2@cornell.edu
Everman	Wesley	North Carolina State University	919-518-0488	wes_everman@ncsu.edu
Farrington	Steven	Gowan Company	334-275-5396	sfarrington@gowanco.com
Fausey	Jason	Valent USA Corporation	419-307-7666	jason.fausey@valent.com
Fidanza	Michael	Penn State University	610-396-6330	maf100@psu.edu
Forney	Raymond	Dupont Crop Protection	302-561-0027	raymond.forney@usa.dupont.com
Gallandt	Eric	University of Maine	207-581-2933	gallandt@maine.edu

Ganske	Donald	DuPont Crop Protection	540-662-6011	donald.d.ganske@usa.dupont.com
Garbett	Mel	Central Garden and Pet	925-948-4000	mgarbett@central.com
Ghantous	Katherine	UMass Cranberry Station	508-295-2212	kghantous@psis.umass.edu
Gianessi	Leonard	CropLife Foundation	202-872-3865	lgianessi@croplifefoundation.org
Gilliam	Charles	Auburn University	334-844-3045	murphan@auburn.edu
Glasgow	Les	Syngenta	336-632-5501	les.glasgow@syngenta.com
Glenn	Scott	University of Maryland	301-405-1331	sglenn@umd.edu
Goddard	Matt	Virginia Tech	540-231-5835	mgoddard@vt.edu
Gomez de Barreda	Diego	Polytechnic University of Valencia	678-662-5003	diegode@btc.upv.es
Gover	Art	Penn State University	814-863-9904	aeg2@psu.edu
Graham	James	Monsanto (Retired)	314-805-7867	jcgrah@charter.net
Graham	Ian	Penn State University	814-863-1014	img102@psu.edu
Graves	Dean	Chevy Chase Club	301-656-6323	deang@chevychaseclub.org
Green	Jerry	Pioneer	484-574-7761	jerry.m.green@pioneer.com
Guiser	Scott	Penn State Cooperative Extension	215-345-3283	sxg6@psu.edu
Gylling	Steve	Gylling Data Management	605-692-4021	staff@gdmdata.com
Hahn	Russell	Cornell University	607-255-1759	rrh4@cornell.edu
Han	Kyung	Penn State University	814-863-1014	kmh373@psu.edu
Hannig	Gregory	Dupont Crop Protection	302-299-7304	greg.hannig@usa.dupont.com
Hanrahan	Richard	Bayer Environmental Science	201-294-5217	rich.hanrahan@bayer.com
Harlow	Chris	North Carolina State University	919-515-8097	chris_harlow@ncsu.edu
Harpster	Tracy	Penn State University	814-865-3190	tlh8@psu.edu
Hearn	Brian	University of Delaware	302-856-7303	bhearn@udel.edu
Hedberg	Rob	USDA National Institute of Agriculture	202-720-5384	rhedberg@nifa.usda.gov
Heimer	Lane	Maryland Department of Agriculture	410-841-5920	laneheimer@gmail.com
Herrick	Robert	FMC Corporation	609-951-3792	bob.herrick@fmc.com
Hess	Dwayne	J.C. Ehrlich Co., Inc.	610-372-9750	dwayne.hess@jcehrlich.com
Hester	Kathleen	Temple University	484-868-2063	hester@aesop.rutgers.edu
Higgins	Simone	Monsanto	314-330-3053	simone.seifert-higgins@monsanto.com
Hines	Thomas	Eastern Shore AREC VPI/SU	757-414-0724	thhines@vt.edu
Hitchner	Erin	Syngenta Crop Protection	609-980-8832	erin.hitchner@syngenta.com
Hodges	Duane	Scotts Company	email him	duane.hodges@scotts.com
Hoffer	Michael	BASF Corp.	919-244-8973	mike.hofer@basf.com
Hooten	Robert	FMC Corporation	816-589-5547	robert_hooten@fmc.com
Horton	Chris	North Carolina State University	919-515-7597	chris_reberg-horton@ncsu.edu
Houseworth	Doug	Arysta LifeScience	904-321-0795	doug.houseworth @arystalifescience.com
Huffman	Janel	Penn State University	814-863-1014	janelhuffman1988@gmail.com
Hutto	Kendall	FMC Professional Solutions	614-392-1384	kendall.hutto@fmc.com
Ikley	Joseph	University of Maryland	410-596-9091	jikley@umd.edu
Isaacs	Mark	University of Delaware	302-856-7303	isaacs@udel.edu
James	J.R.	Syngenta Crop Protection	336-632-6000	j_r.james@syngenta.com
Jemison	John	University of Maine - Cooperative Extension	207-581-3241	jjemison@umext.maine.edu

Jester	Jen	Virginia Tech	540-231-5835	jesterj@vt.edu
John	Wynn	Dupont Crop Protection	302-451-5835	w-wynn.john@usa.dupont.com
Johnson	Jon	Penn State University	814-863-1184	jmj5@psu.edu
Johnson	Quintin	University of Delaware	302-856-2585	quintin@udel.edu
Johnson	David	Pioneer Research	515-727-7234	david.h.johnson@pioneer.com
Johnson	Roy	Waldrum Specialties Inc	215-817-0637	rjoh834880@aol.com
Jones	Patrick	University of Tennessee	865-974-6730	patajone@utk.edu
Jordan	Grant	A.C.D.S. Research Inc.	315-587-2140	gljordan@acdsresearch.com
Jotcham	Jim	Marbicon Inc.	902-538-7101	marbicon@eastlink.ca
Kahl	Jerry	J.C. Ehrlich Co., Inc.	610-372-9750	jerry.kahl@jcehrlich.com
Kalmowitz	Kathie	BASF Corp.	919-270-4592	kathie.kalmowitz@basf.com
Kaminski	John	Penn State University	814-865-3007	kaminksi@psu.edu
Kammerer	Melanie	Penn State University	814-863-1014	
Kao-Kniffin	Jenny	Cornell University	607-255-8886	jtk57@cornell.edu
Keene	Clair	Penn State University	814-863-1014	clk5285@psu.edu
Keese	Renee	BASF Corp.	919-547-2791	renee.keese@basf.com
Kelly	Patrick	Anne Arundel County Recreation and Parks Dept.	443-262-9655	rpkell96@aacounty.org
Kirfman	Gary	Valent USA Corporation	616-560-2510	gkirf@valent.com
Kleczewski	Victoria	DuPont	219-208-9499	victoria.a.kleczewski@dupont.com
Knight	Alexandra	North Carolina State University		amknigh4@ncsu.edu
Koo	Suk-Jin	Moghu Research Center		moghu@moghu.com
Kordbacheh	Farnaz	Cornell University	410-830-0781	fk226@cornell.edu
Kunkel	Dan	Rutgers - IR-4 Headquarters	732-932-9575	kunkel@aesop.rutgers.edu
Kyde	Kerrie	MD Dept. of Natural Resources	301-948-8243	kkyde@dnr.state.md.us
Larson	Steven	BASF Corp.	281-684-4542	steven.larson@basf.com
Leahey	Caitlin	Cornell University	607-255-4747	caitlin.leahey@gmail.com
Leather	Gerald	WVU Cooperative Extension	304-822-5013	gerald.leather@mail.wvu.edu
Lenker	Jessica	PA Department of Agriculture	717-787-7204	jeslenker@pa.gov
Lightfoot	Rachel	Crop Management Strategies	610-767-1944	ral_cms@fast.net
Linde	Douglas	Delaware Valley College	215-489-2260	douglas.linde@delval.edu
Lindquist	John	University of Nebraska	402-472-2771	jlindquist1@unl.edu
Lingenfelter	Dwight	Penn State University	814-865-2242	dwight@psu.edu
Little	Neith	Cornell University	607-346-5958	ngl7@cornell.edu
Little	Daniel	North Carolina State University	517-974-3000	dalittle@ncsu.edu
Lloyd	Kirsten	Penn State University	814-863-1184	kll24@psu.edu
Loecke	David	PBI Gordon	816-547-8373	vcline@pbigordon.com
Lohmann	Henry	Nicholls University (Retired)	631-286-1078	halohmann@aol.com
Lough	Katherine	University of Maine	207-581-2923	klough@maine.edu
Loughner	Dan	Dow AgroSciences LLC	609-921-1856	dloughner@Dow.com
Lurvey	Edith	Cornell University -NYSAES	315-787-2308	ell10@cornell.edu
Lycan	Darren	Syngenta Lawn and Garden	315-635-2818	darren.lycan@syngenta.com
Lydon	John	USDA-ARS	301-504-6470	john.lydon@ars.usda.gov
Lyman	Betsy	National Park Service	570-588-0513	betsy_lyman@nps.gov
Mahoney	Matthew	Bayer CropScience	410-822-5215	matt.mahoney@bayer.com
Majek	Bradley	Rutgers University	856-455-3100	majek@aesop.rutgers.edu
Manley	Brian	Syngenta Biotechnology, Inc.	919-281-2162	brian.manley@syngenta.com

Mansue	Carrie	Rutgers University	732-932-9711 x 116	cmansue@aesop.rutgers.edu
Marble	Chris	Auburn University		marblsc@auburn.edu
Marose	Betty	USDA-ARS	301-504-5156	bhmarose@comcast.net
Marschner	Caroline	Cornell University	360-915-4778	cam369@cornell.edu
Mathers	Hannah	Ohio State University	614-247-6195	mathers.7@osu.edu
Mathews	Sudeep	University of Maryland	410-228-8800	samathew@umd.edu
Matthews	Michael	For-Shore Weed Control	609-693-6999	mike@for-shore.com
Mayonado	David	Monsanto Company	410-726-4222	david.j.mayonado@monsanto.com
McCullough	Patrick	University of Georgia	770-228-7276	pmccull@uga.edu
McDonald	Steven	Turfgrass Disease Solutions LLC	610-633-1878	turfgrassdiseasesolutions @yahoo.com
McDonnell	Brian	National Park Service	570-588-0534	brian_mcdonnell@nps.gov
McNulty	Brendon	Virginia Tech	540-231-5807	bmcnulty@vt.edu
Meck	Elijah	Syngenta	772-567-5218	elijah.meck@syngenta.com
Mervosh	Todd L.	Connecticut Ag. Exp Station	860-683-4984	todd.mervosh@ct.gov
Meyer	Tanya	Sinnemahoning Invasive Plant Mgmt. Area	814-486-1740	tanya@sipmapa.org
Milbrath	Lindsey	USDA-ARS Robert W. Holley Center	607-254-7268	lindsey.milbrath@ars.usda.gov
Miller	Raymond	Dow AgroSciences LLC	813-363-9059	rcmiller2@dow.com
Mirsky	Steven	USDA-ARS ANRI SASL	301-504-7199	steven.mirsky@ars.usda.gov
Mohler	Charles	Cornell University	607-255-0199	clm11@cornell.edu
Morris	Tom	Univ. of Connecticut-SARE	860-486-0637	thomas.morris@uconn.edu
Mortensen	David	Penn State University	814-865-6679	dmortensen@psu.edu
Mosdell	Dean	Syngenta Crop Protection	805-480-0514	dean.mosdell@syngenta.com
Munsterman	Chris	Syngenta Crop Protection	304-261-9564	chris.munsterman@syngenta.com
Myers	Donald D.	Bayer Crop Science	919-549 2529	don.myers@bayer.com
Naedel	Matthew	Penn State University	814-863-1613	mbn112@psu.edu
Neal	Joseph	North Carolina State University	919-515-9379	joe_neal@ncsu.edu
Norton	Lawrence	Bayer Environmental Science	610-814-6220	larry.norton @bayercropscience.com
O'Barr	John H.	BASF Corp.	717-386-8259	john.obarr@basf.com
O'Connell	James	UMass Cranberry Station	508-295-2212	jameso@umext.umass.edu
O'Neal	William	AMVAC Chemical Corporation	919-619-3095	boneal@sprynet.com
O'Neill	Brian	Weeds Inc	610-358-9430	bgon58@aol.com
O'Neill	Drew	Weeds Inc	610-358-9430	weeds@weedsinc.com
Olson	Brian	Dow AgroSciences LLC	317-337-7245	bdolson@dow.com
Orlowski	John	University of Kentucky	518-596-9311	john.orlowski@uky.edu
Orr	James	Asplundh	800-248-8733	jimorr@asplundh.com
Pacchioli	Marc	Crop Management Strategies	610-767-1944	cms1@fast.net
Palmer	Cristi	Rutgers - IR-4 Headquarters	732-932-9575	palmer@aesop.rutgers.edu
Palmer	Butch	Reality Research	315-945-0945	whpalmer@localnet.com
Pannill	Philip	U.S. Fish and Wildlife	304-876-7432	phil_pannill@fws.gov
Parker	Astrid	Bayer Environmental Science	919-625-3350	astrid.parker@bayer.com
Parochetti	James	USDA-NIFA (retired)	202-401-4354	jamesparochetti@aol.com
Parrish	Jason	Ohio State University	440-225-4831	parrish.174@osu.edu
Patches	Kelly	Penn State University	814-863-1014	kmp5212@psu.edu

Patton	Aaron	Purdue University	765-494-9737	ajpatton@purdue.edu
Pawlak	John	Valent USA Corporation	925-256-2700	jpawlak@valent.com
Pennucci	Annamarie	Northeast T&O Research	610-895-8480	aapennucci@yahoo.com
Phillips	Jeffery	Monsanto	330-402-2591	jeffrey.i.phillips@monsanto.com
Phillips	Bill	US EPA, OPP	703-308-8099	phillips.bill@epa.gov
Pieczarka	David	Gowan Company		Pieczarka@gowanco.com
Poffenbarger	Hannah	USDA-ARS ANRI SASL	301-504-7199	hpoffenb@gmail.com
Post	Angela	Virginia Tech	540-231-5807	arpost@vt.edu
Prostak	Randy	University of Massachusetts	413-577-1738	rprostak@umext.umass.edu
Pyle	Steve	Syngenta Crop Protection	336-632-2236	steve.pyle@syngenta.com
Qin	Zhong	South China Agriculture Univ.	607-254-5365	q_breeze@126.com
Rana	Aman	Virginia Tech	757-237-3412	ranaaman29@gmail.com
Rardon	Patrick	Dupont Crop Protection	302-366-5546	patrick.l.rardon@usa.dupont.com
Ratnayake	Sunil	US EPA	703-308-8191	ratnayake.sunil@epa.gov
Reeb	Bryan	The Ohio State University	740-225-2014	reeb.22@osu.edu
Reed	Joseph	FMC Corportion	217-649-3249	joseph.reed@fmc.com
Richardson	Robert	North Carolina State University	919-515-5653	rob_richardson@ncsu.edu
Richtmyer III	Richard	Cornell University	607-342-1695	rjr39@cornell.edu
Riffle	Michael	Valent USA Corporation		mriff@valent.com
Ritter	Ronald	University of Maryland	301-405-1329	rlritter@umd.edu
Rogers	Gregory	DuPont Crop Protection	443-309-0148	gregory.s.rogers@usa.dupont.com
Rorem	Kent	North Carolina Department of Agriculture	910-675-2314	kent.rorem@ncagr.gov
Roten	Rory	North Carolina State University	919-610-6165	rory_roten@ncsu.edu
Rupp	Peter	Maryland Department of Agriculture	410-841-5920	halletdc@mda.state.md.us
Ryan	Matthew	Cornell University	814-574-3028	mrr203@psu.edu
Sandler	Hilary	UMass Cranberry Station	508-295-2212 x21	hsandler@umext.umass.edu
Saunders	David	DuPont Crop Protection	515-334-4485	david.w.saunders@usa.dupont.com
Schou	Bert	ACRES Research	304-456-5558	bertschou@aol.com
Scoresby	Rene	Green Light	715-298-3315	renescoresby@yahoo.com
Scott	Barbara	University of Delaware	302-856-2585	bascott@udel.edu
Sellman	Leroy	Maryland Department of Agriculture	410-841-5920	halletdc@mda.state.md.us
Senesac	Andrew	Cornell University Cooperative Extension	631-727-3595	afs2@cornell.edu
Serensits	Thomas	Penn State University	610-360-5985	tjs204@psu.edu
Sharma	Shiv	FMC Corporation	215-299-6871	shiv.sharma@fmc.com
Shirtliffe	Steve	University of Saskatchewan	306-966-4959	steve.shirtliffe@usask.ca
Smith	Richard	University of New Hampshire	603-862-2724	richard.smith@unh.edu
Smith	Mark	MD Department of Agriculture	410-841-5920	smithmj@mda.state.md.us
Smith	Adam	Virginia Tech	540-231-5835	urzsmith@vt.edu
Smith	Larissa	Virginia Tech	540-231-6323	lls14@vt.edu
Smith	Chris	Bluegrass Lawn Service and Landscaping	215-364-7000	Service@BGLawnService.com
Snyder	Eli	Penn State University	814-863-1014	elina.snyder@gmail.com
Spak	David	Bayer Environmental Science	919-625-3350	david.spak@bayer.com
Stachowski	Paul	Cornell University	315-364-8296	pjs16@cornell.edu

Stalter	Richard	St. John's University	718-990-6288	stalterr@st.johns.edu
Steffel	James	LABServices	610-562-5055	jim@labservices.com
Swanton	Clarence	University of Guelph	519-824-4120 x 53392	cswanton@uoguelph.ca
Sweet	Robert	Cornell University (Emeritus)	607-273-7106	mw45@cornell.edu
Szylvian	Andrea	US EPA, Region 1 Pesticide Program	617-918-1198	szylvian.andrea@epa.gov
Tan	Siyuan	BASF Corp.	919-547-2679	siyuan.tan@basf.com
Tardif	Francois	University of Guelph	519-824-4120 x 53395	ftardif@uoguelph.ca
Taylor	Erin	Michigan State University	517-355-0271	hiller12@msu.edu
Taylorson	Raymond	USDA & URI (Retired)	410- 778-9088	raymondtaylorson@verizon.net
Teasdale	John	USDA-ARS	301-504-5504	john.teasdale@ars.usda.gov
Tekiela	Dan	Virginia Tech	540-231-6323	tekiela2@vt.edu
Thomas	Gar	BASF Corp.	410-885-5920	garfield.thomas@basf.com
Thomas	W. Joseph	BASF Corp.	919-547-2549	walter.e.thomas@basf.com
Toressen	Kristine	Cornell University	518-596-9311	kst49@cornell.edu
Tranel	Patrick	University of Illinois	217-333-1531	tranel@illinois.edu
True	Sarah	North Carolina State University	919-812-5679	sltrue@ncsu.edu
Vail	Gordon	Syngenta Crop Protection	336-632-5596	gordon.vail@syngenta.com
Vanderwerker	Cindy	National Park Service	518-664-9821	cindy_vanderwerker@nps.gov
VanGessel	Mark	University of Delaware	302-856-7303	mjv@udel.edu
Vanini	J.Tim	New Dimensions Turf	716-319-7495	tim@ndturf.com
VanWychen	Lee	WSSA Liason	202-746-4686	Lee.vanWychen@wssa.net
Vargas	J.Javier	University of Tennessee	865-974-7324	jvargas@utk.edu
Vea	Elymar V.	IR-4 Headquarters	410-923-4880	evvea@comcast.net
Venner	Katelyn	Virginia Tech	540-231-5835	katevenn@vt.edu
Vitolo	David	Syngenta Crop Protection	916-316-6951	david.vitolo@syngenta.com
Walls	Bobby	FMC Corporation	919-735-3862	bobby.walls@fmc.com
Wells	M. Scott	North Carolina State University	919-741-9876	mswells@ncsu.edu
Welterlen	Mark	PBI/Gordon Corporation	816-460-6205	mwelterlen@pbigordon.com
Wenskus	Tim	NYC Parks and Recreation	212-360-1427	tim.wenskus@parks.nyc.gov
West	Amanda	North Carolina State University	919-812-5679	amwest@ncsu.edu
White	Linda	National Park Service	518-664-9821	linda_white@nps.gov
White	Tim	Crop Management Strategies	610-767-1944	cms1@fast.net
Whitehouse	Stephanie	Cornell University	607-255-4747	sew54@cornell.edu
Willis	John	Monsanto Company	937-418-5667	john.b.willis@monsanto.com
Wilson	Samuel	FMC Corportion	803-413-7824	Sam.Wilson@fmc.com
Wilson	Henry	Virginia Tech	757-414-0724	hwilson@vt.edu
Wollam	John	Bayer Crop Science	816 242-2000	john.wollam@bayer.com
Yarborough	David	University of Maine	207-581-2923	davidy@maine.edu
Zawierucha	Joe	BASF Corp.	919-547-2095	joseph.zawierucha@basf.com
Zelna	Jeffrey	Syngenta Crop Protection	315-498-4259	jeff.zelna@syngenta.com
Zhang	Jia-en	South China Agricultural Univ.		jeanzh@scau.edu.cn

AUTHOR INDEX
(by Abstract Number)

KEYWORD INDEX
(by Abstract Number)

Northeastern Weed Science Society (NEWSS)
P.O. Box 25
Woodstown, NJ 08098
USA

ISBN 978-1-5108-3340-1

CURRAN ASSOCIATES INC.
proceedings.com